MW01618479

75 years of innovation

Kennard C. Braband
Bernard J. Smith

> Assembly area in Cedar Rapids, Iowa, in 1973. The lines from the ceiling supplied electricity and compressed air to the work stations.

Rockwell Collins

75 years of innovation

Creative Direction Duane Wood
Design/Art Direction Eric Johnson
Editor Elinor Day

All rights reserved. No part of this publication may be reproduced, stored in a retrieval system, or transmitted, in any form or by any means, electronic, mechanical, photocopying, recording, or otherwise, without prior written permission from the publisher and Rockwell Collins.

Copyright © 2010 Rockwell Collins, Inc.

First published in the United States of America by
WDG Communications Inc.
1615 32nd Street NE
Suite 2
Cedar Rapids, Iowa 52402-4072
Telephone (319) 396-1401
Facsimile (319) 396-1647

Library of Congress Cataloging-in-Publication Data

Braband, Ken C.
Rockwell Collins : 75 years of innovation / Kennard C. Braband, Bernard J. Smith.
p. cm.
Includes bibliographical references.
ISBN 978-0-9826138-1-8 (alk. paper)
1. Rockwell Collins (Firm)--History. 2. Radio supplies industry--United States--History.
3. Telecommunication equipment industry--United States--History. 4. Electronic industries--United States--History. I. Smith, Bernie, 1953- II. Title.
HD9696.R364C6424 2010
338.7'621380973--dc22

2010015103

Printed in the United States of America

10 9 8 7 6 5 4 3 2 1

EMERGENCY
OFF

Table of contents

‹ Collins Radio employees conduct tests on the Apollo command module communication and data system equipment.

Introduction

Any enterprise that can trace its roots 75 years into the past must have many stories to tell. That is particularly true of Rockwell Collins. The nature of its industry and the far-reaching impact of communication and aviation technology have tied the company inextricably to global events and to world-altering cultural and societal transitions.

Rockwell Collins has been shaped by history and at times it has influenced history. Its people and products have played key roles in great explorations of the Earth's poles, in mankind's ventures into space, and in the evolution of air travel and warfare. Today, that tradition of innovation lives on through advanced technology and customer-focused solutions that are shaping the future of communication and aviation electronics.

The company's first 50 years were marked by great growth and numerous successes, and by the eventual acquisition of Collins Radio by Rockwell International. The stories of those years and transitions were documented in the book titled "The First 50 Years" by Ken C. Braband. That text and many of the photographs and illustrations are reproduced in this new volume, which commemorates the 75th anniversary of the company's founding.

In the most recent 25 years of the company's history, the pace of change quickened. When the previous volume was published, personal computers were young and software was a new and growing frontier. GPS technology, which now seems to be everywhere, was in its infancy. Ronald Reagan was president and defense spending was growing. The Soviet Union still loomed ominously just a few short years before it dissolved. The Berlin Wall had yet to come down, and commercial aviation was taking off.

These were just a few of the events that would help shape Rockwell Collins into what it is today: a company focused on its customers, its people, its far-reaching community, and the future. ▪

‹ Arthur Collins — a young entrepreneur — oftentimes mixed business meetings with life's normal occurrences. According to his children, Collins made every attempt to leave the office in time to be home for dinner with his family.

Radio Mechanics
25¢
Edited by M. B. Sleeper
November, 1926
Blueprint Section Every Month
RADIO AGE
The Magazine of the Hour
25 CENTS
APRIL. 1925
Age
Radio Magazine
Popular
25¢
with which is combined The
FEBRUARY 1926
How to Build the New Orthophase Radio Receiver

A young radio enthusiast

Early amateur radio operators were mainly hobbyists, but there was a sense of discovery during the infancy of radio that provided something more. Radio was the new thing, comparable to what computers mean to technological whizzes in the 1980s. And like the computer hobbyists of today who are writing their own programs and building their own equipment, amateur radio operators in the 1920s were contributing to the knowledge of practical aspects of radio art.

‹ The variety of radio magazines testified to the popularity of the hobby in the 1920s.

One person caught up in the excitement of radio was Arthur Andrew Collins.

Born in Kingfisher, Oklahoma, on September 9, 1909, Collins moved to Cedar Rapids, Iowa, at an early age when his father, Merle (or M. H. as he preferred to see it written), established The Collins Farms Company there.

With the Collins Farms Company, M. H. Collins brought new ideas to the stoic profession of farming. The elder Collins reasoned that the efforts of scientists and engineers could do for farming what they did for nearly every other industry in America.

"Why not manufacture food for the American consumer as cheaply as motor cars and radios are manufactured?" M. H. asked in a publication which explained his new ideas. "Why not produce food on a large scale by intensive farming methods in Iowa, where high yields could be obtained and at a low cost?"

The primary object of the farm company was to produce grain at low cost. M. H. Collins felt that too many farmers mixed grain production with livestock raising, and as a result, both were unprofitable.

He implemented his plans by convincing landowners he could improve the profitability of their tenant farms. Farms of 160 to 320 acres were planted to a single crop and were rotated as a single tract in a unit group of farms, each embracing 1,500 to 2,000 acres. Each unit of contiguous farms was put under the supervision of a salaried foreman who directed the tractor operators. The most modern machinery was employed for every operation — four-row cultivators, rotary hoes, deep disc plows, two-row corn pickers, fast trucks for marketing crops, and semi-trailers for moving machinery. For maximum use of all this new machinery, electric lights were placed on the tractors, allowing round-the-clock operation. Other practices initiated by

> One of the modern implements used by the Collins Farm Company was a "cultipacker" which pulverized and firmed the seed bed. Note the electric lights on the tractor, another innovation.

The Collins Farms Company included installing drainage tile, erecting fire-proof ventilated grain storage bins, and using legumes to replace nitrogen in the soil.

At its peak the company operated 60,000 acres of farmland in 31 Iowa counties, with wealthy businessman M. H. Collins at the helm of the corporation.

At about the age of nine, Arthur Collins became deeply interested in the new marvel of radio, although at first M. H. apparently did not think highly of his son's tinkerings with radio. Arthur and another early boyhood radio devotee, Merrill Lund, made their first crystal receivers at the Lund home at 1644 D Avenue in Cedar Rapids. The sets used variable condensers inside a tube. Merrill's father worked in the tube department at Quaker Oats Co. and made tubes of the size the boys needed. Using thumb tacks for contact points, they wrapped wire around the tubes. From iron plates they fashioned their own transformers, and rigged a 60-foot spark antenna with a lead-in through a basement window of the Lund home. Merrill's father asked them to find another location for their equipment after lightning struck the radio set and blew it up.

Arthur brought over two coaster wagons and the two boys transported the damaged equipment to the Collins home at 1725 Grande Avenue. Although M. H. did not approve of the mess it was going to cause, Arthur hauled the equipment to his room once his father was out of sight.

"I used a Quaker Oats box to wind the tuning coil and used a Model T spark coil," he told a *New York Times* reporter in 1962. "The main piece of the station's machinery was the transmitter. Other parts of the station were recruited from a rural telephone service. The way we calibrated was to pick up signals from WWV (the Navy's station in Arlington, Virginia)."

Arthur also used pieces of coal or coke for a rectifier, glass towel racks for insulators and a toy motor. Those early efforts reflected a lot of experimenting that led to successively more reliable, higher-performance radios.

Another boyhood friend in Cedar Rapids who also had an interest in radio was Clair Miller.

"Arthur had big expensive tubes as a kid while all the rest of us had were peanut tubes," Miller told a reporter in 1965.

The article quoted another neighbor's recollection of early days in the Collins family neighborhood: "We sensed that Arthur was different, but we did not know that he was a genius. When the rest of us were out playing cowboy and Indian, Arthur was in the house working on his radios."

One day Arthur's mother invited the neighborhood boys into the Collins home. "I think she did so because she wanted us to realize that Arthur was different from the rest of us. We went upstairs to see what he was doing. He had a room that overlooked the yard. It was loaded with radio stuff. We knew a little about radio. We had been playing around with crystal sets ourselves. But Arthur had one wall covered with dials and switches, everything under the sun."

The Federal Radio Commission, the predecessor to the Federal Communications Commission, passed a radio act whereby amateurs could get licenses. Arthur took the test and got his license in 1923 at the age of 14.

As M. H. Collins recognized his son's talent and ambition toward radio, he looked for ways to help with Arthur's hobby. In about 1924, Arthur's father purchased a new tube costing $135 and other high voltage equipment.

"When I was a youngster there were two real active amateurs (in Cedar Rapids)," Arthur recalled. "One was Henry Nemec and the other was Clark Chandler." Collins and Leo Hruska, another friend who had constructed a crystal receiver, used to receive the stations of Nemec and Chandler, and considered their talks with the more experienced radio operators quite an achievement.

Nemec recalled how he first met the young boy with the extensive radio knowledge. M. H. Collins had asked Nemec to meet with his son so Nemec could teach Arthur some of what he knew about amateur radio, "But there wasn't much that he didn't already know," Nemec said.

In a 1978 interview for an article about Henry Nemec in the *Cedar Rapids Gazette*, Arthur Collins recalled the early days when he purchased a vacuum tube from Nemec, and several years later when Nemec, who worked for the police department, and another patrolman, Frank Bukacek, parked their squad car in front of Collins' house while the three were inside talking radio. Collins said he, Nemec and Bukacek got together so frequently, and the squad car was parked in front of the house so often, that neighbors began to wonder whether Collins was in some kind of trouble with the police.

At the time there was little formal instruction in the science of radio. Several two-day short courses were given at Iowa State College at Ames, and Arthur is said to have attended the first of these while still wearing knickers. Carl Mentzer later sponsored a course in radio at the University of Iowa. These courses, along with several periodicals, including *Wireless Age* and *QST*, comprised most of the current radio knowledge of the time, other than word-of-mouth information.

By the time he was a teenager, Arthur had constructed an amateur radio station using purchased components, makeshift materials and his own ingenuity. Arthur's family had moved to a new home at 514 Fairview Drive, and his equipment moved with him.

By the age of 15, Collins had communicated with other amateur "hams" in the United States and many foreign countries. The custom of exchanging postcards after

› This highly-retouched photograph of 15-year-old Arthur Collins originally appeared in the August 11, 1925 *Cedar Rapids Gazette*. The sloping walls of the attic at 514 Fairview Drive held many of the QSL postcards Collins received from other "ham" radio enthusiasts.

> A young Arthur Collins — The Collins Radio Company founder, age 15.

a contact was made had already been established in the amateur world, and one wall of Collins' attic room was covered with so-called QSL cards.

One card from Australia came from a radio operator who regarded America's prohibition of alcohol as a joke.

"How does it feel to stay sober?" were the words the Australian ham wrote to 15-year-old Arthur.

And during a contact with a person in Chile, the South American operator asked to be excused from the radio conversation because a volcano was erupting and interfering with the talk. "He referred to it as if it was in his backyard," Collins told a *Cedar Rapids Gazette* reporter in 1925.

The reporter, Gladys Arne, had gone to the Collins home to talk with the 15-year-old boy because he had made a radio contact that put him on the front pages of newspapers all over the country.

During the winter of 1924-25, Collins had become familiar with John Reinartz, a 31-year-old German immigrant who was prominent in radio circles because he developed a "tuner" or receiver capable of predictable selectivity and reception. Reinartz had authored several articles on the subject for radio magazines. Reinartz and Collins carried on experiments, particularly in the use of short wavelengths.

Because of Reinartz's radio success, he was chosen as the radio operator for a scientific expedition to the continent of Greenland. The MacMillan expedition set sail from the coast of Maine on the ships *Bowdoin* and *Perry* in early 1925. One of the explorers was U.S. Navy Lt. Cdr. Richard E. Byrd.

The plan was for the *Bowdoin* to make daily radio reports to the U.S. Naval radio station, but because of atmospheric problems, the land station in Washington, D.C., was unable to consistently receive Reinartz's messages.

Then word spread that a 15-year-old boy in Cedar Rapids had made contact with the expedition. Throughout the summer of 1925, Arthur Collins accomplished a task that even the U.S. Navy found difficult. Using a ham radio that he himself had built, he talked by code with Reinartz in Greenland night after night. His signals reached the expedition more clearly than any other. After each broadcast, young Collins took the messages from the expedition down to the Cedar Rapids telegraph office and relayed to Washington the scientific findings that the exploratory group had uncovered that day.

Collins' exclusive contact with the expedition soon became a nationwide news story that won him acclaim as a radio wizard. The August 4, 1925 *Cedar Rapids Gazette* told the story:

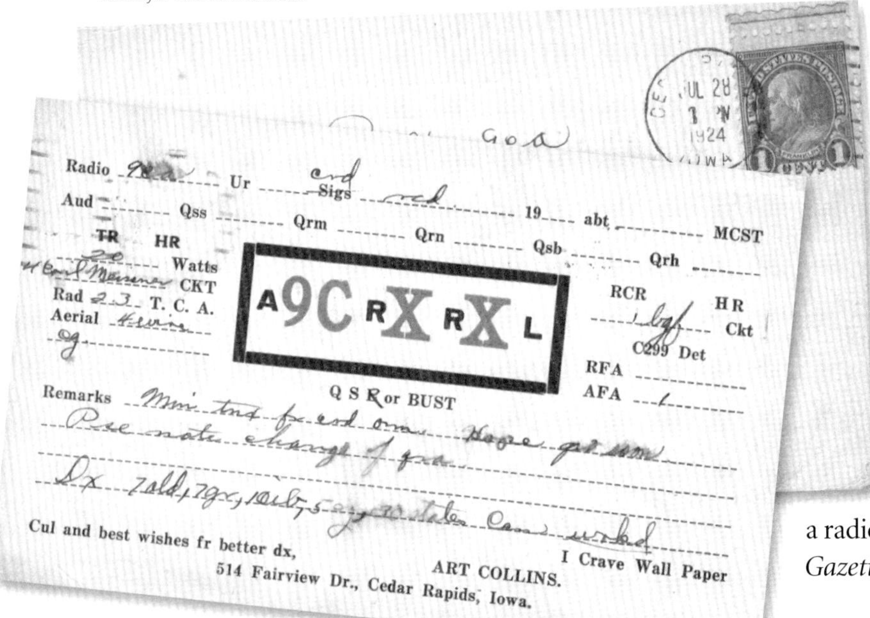

> An original QSL card from Arthur Collins that was sent as a confirmation of contact in 1924.

Courtesy of the G4UZN Collection.

> Before departing for the southwest, the operators of 9ZZA posed with their mobile shortwave station. From left: Winfield Salisbury, Paul Engle, and Arthur Collins.

"The mysterious forces of air leaped the boundary of thousand of miles to bring Cedar Rapids in touch with the celebrated MacMillan scientific expedition at Etah, Greenland, and wrote a new chapter into the history of radio. Sunday, Arthur Collins, 514 Fairview Drive, 15-year-old radio wizard, picked up the message from the expedition's ship *Bowdoin*, at twenty meters (wavelength), at about 3 o'clock and conversed in continental code for more than one hour. It was the first time the expedition and any United States radio station had communicated at that wavelength. Messages were received by Collins for the National Geographic Society, which is sponsoring the expedition, and for others, and were sent out from here by telegraph. Arthur Collins is the son of Mr. and Mrs. M. H. Collins and is a student at Washington High School. He has been a radio fan for years, and has himself constructed most of his apparatus. His equipment is in a small room on the third floor of the Collins home. His station is known as 9CXX. The local boy told a *Gazette* reporter today that although he had been in wireless communication with Australia, Scotland, England, India, Puerto Rico, Guam, and Mexico, he never had received a greater thrill than when he talked to his friend on the famous expedition bound northward to explore a mystic continent."

One week later, a follow-up article in the *Gazette* concluded: "Though only 15, he is true to his trust. For he hopes to realize great radio ambitions, by and by."

At the age of 16, Collins was asked to write a technical article for *Radio Age* which was published in the May, 1926, issue. One statement in that article foreshadowed the motivational force which was to lead him to "great radio ambitions."

"The real thrill in amateur work comes not from talking to stations in distant lands … but from knowing that by careful and painstaking work and by diligent and systematic study you have been able to accomplish some feat, or establish some fact that is a new step toward more perfect communication."

Arthur's reputation in the radio world grew. Radio operators around the country who had heard about his contacts with the MacMillan expedition wrote to him to ask how he did it.

Collins continued his electronics education by taking courses at Amherst College in Massachusetts, Coe College in Cedar Rapids, and the University of Iowa in Iowa City.

In 1927, he and two friends organized an expedition of sorts of their own. Collins, Paul Engle, and Winfield Salisbury outfitted a truck with short wave transmitting and receiving equipment and took a summer trip to the southwest states. Using power of 10 watts they conducted experiments in connection with the U.S. Naval Observatory in Washington, D.C. Leo Hruska stayed behind in Cedar Rapids to operate the base station for the study.

Like Collins, both Engle and Salisbury would later go on to achieve recognition in their particular chosen fields — Engle as a poet and professor at the University of Iowa, and Salisbury as a noted physicist who would make significant contributions to studies initiated by Collins. ▪

> A U.S. Postage Stamp issued in 1964 commemorating amateur radio.

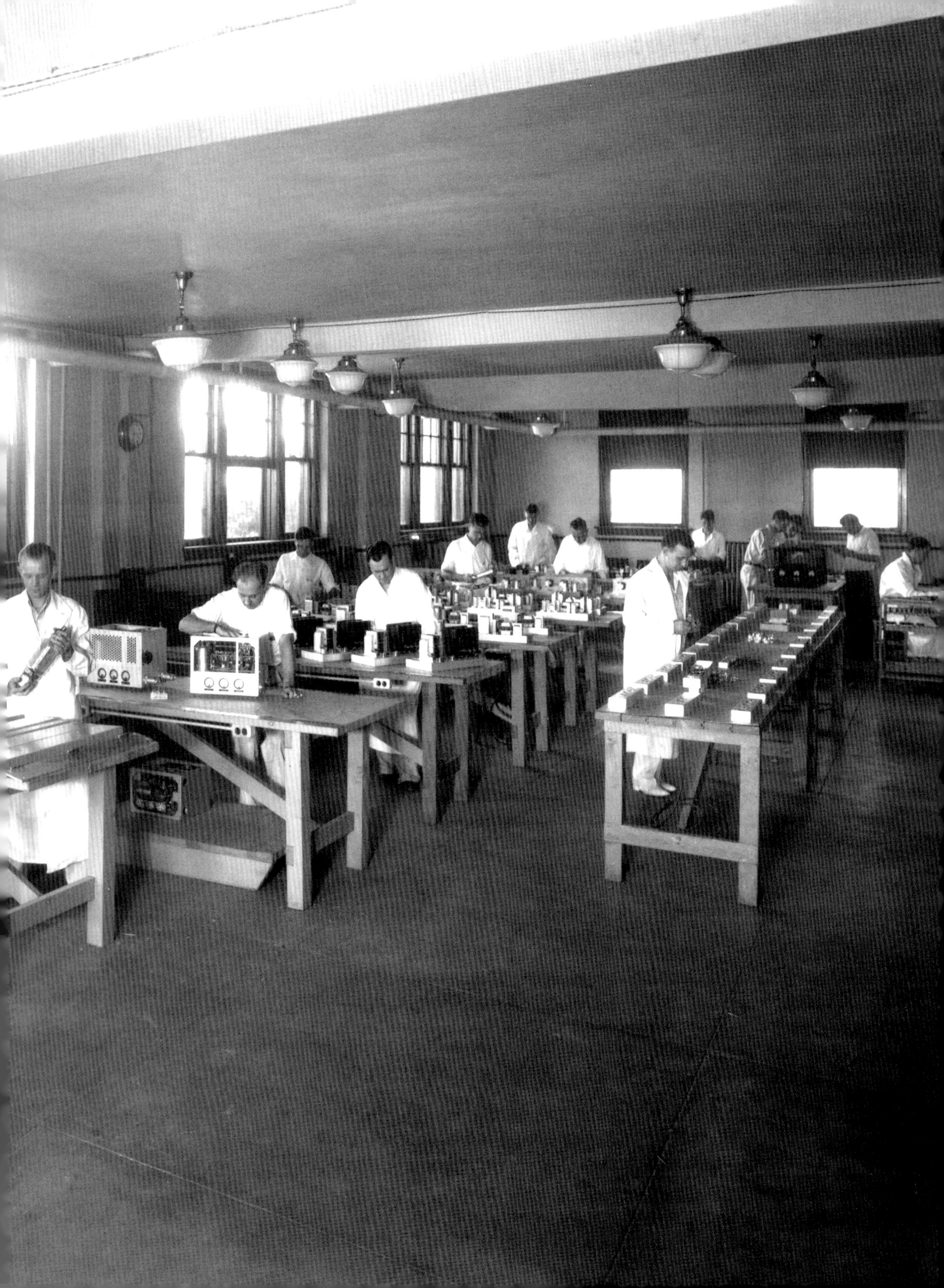

Up from the cellar

In 1930, Collins married Margaret Van Dyke. By the end of 1931 he had set up a shop in the basement of their home at 1620 6th Avenue S.E., previously the home of his grandparents. Arthur began to produce transmitters to order.

When the depression hit with full force in 1931, 23-year-old Collins turned his hobby into a vocation.

"I picked what I was interested in," he told *Forbes* magazine years later, "and looked for a way to make a living."

‹ The assembly area at 2920 First Avenue as it appeared in 1934. Workers were identified, from left, as: Elmer Koehn, Don Anderson, Don Holmes, Don Wheaton, Roy Olson, John Dayhoff, Arlo Goodyear, Don Cole, Don Andersen, R. E. Samuelson, Walt Wirkler, Leonard Braun, and Gerald Ozburn.

This was the first time radio transmitting apparatus, of any power output, was available for purchase as an assembled and working unit. In fact, components were hard to come by; they varied widely in characteristics, and there was little, if any, pattern to their construction. Most hams had their radio equipment scattered around a room, usually in a basement or attic where the sight of tubes and wires wouldn't clutter up living areas of a home. Their equipment was strictly functional, almost to the point of inefficiency.

Collins' ham gear was designed to eliminate the clutter by packaging the equipment in neat units. The concept proved that correctly engineered construction not only stabilized the circuitry but also made its behavior predictable. Collins designed circuits, fabricated chassis, mounted and wired in components, tested, packed and shipped each unit. Because the gear was precisely engineered and well-built with the best parts available, it gave years of trouble-free service.

A later article in the *New York Times* quoted a ham as saying, "Collins brought us up from the cellar and put us into the living room." The industrial philosophy of Collins products — quality — was established at the very start.

The first advertisement for this new line of products appeared in the January, 1932 issue of *QST*, with the firm name given as Arthur A. Collins. Two issues later, in March, 1932, the firm name appeared as Collins Radio Transmitters with Arthur's name and call number below. Both notices were two-inch advertisements, but by May the size was increased to six inches. In October the first full-page ad appeared and by December, the firm's name was listed as Collins Radio Company.

Long-time friend Jiggs Ozburn recalled his first meeting with Collins.

"I first met Arthur Collins at a ham club meeting at his house (factory in

> The basement of this home at 1620 6th Avenue SE in Cedar Rapids, Iowa, was the first factory in which Art Collins produced radio transmitters to order.

basement). I hadn't finished high school and he hadn't finished college (he never did). Art had been making ham transmitters for about a year. He was tall and slender and very quiet. He came from a well-to-do family whose fortune was taking a beating in those depression years, so Art was pretty much on his own."

The Great Depression, which began after the stock market crash in 1929, was having a devastating effect on the Collins Farms Company. In 1931, M. H. Collins sold the firm to an east coast insurance company.

Arthur originally started his company as sole owner with only one employee, Clair Miller, who had just been graduated from Iowa State College. But as his business grew, he added personnel, including some who came from his father's farm company. Among them were John Dayhoff and Ted Saxon.

Orders came in and the company grew. In 1933, Collins Radio Company moved out of the basement factory and into leased space at 2920 First Avenue in Cedar Rapids, now headquarters for the local Salvation Army.

Business in general in 1933 was not good, to put it mildly, but radio had come of age, and Collins recognized the need for advancement in the radio communications field.

One Saturday morning that year, Collins telephoned Arlo Goodyear and offered him a job for two months if he was willing to work on Sundays. Out of work for months and with a wife and baby, Goodyear jumped at the opportunity.

"There it was in the middle of the Depression and he was asking me if I minded working on Sunday," Goodyear later wrote. "I would have worked on Shrove Tuesday."

Collins and his work force of one arrived at the building only to discover that neither had a key to get into the basement area, where the company was to begin production the next morning.

Collins looked at the door, looked at Goodyear, and said, "Well, we've got to get going. Catch me so I won't fall on my head." Collins charged the locked door and the new plant got underway with a bang and a shattered front door.

> One of the earliest advertisements placed by Arthur Collins appeared in the September, 1932 issue of *QST*, a magazine for amateur radio hobbyists.

A manual is now available describing COLLINS transmitters, speech equipment, condenser microphones and transformers, complete with full specifications and circuit diagrams.

Send 25c in coin

COLLINS RADIO TRANSMITTERS
W9CXX, CEDAR RAPIDS, IOWA

> The basement of this building at 2920 First Avenue became the factory and office area for Collins Radio Co. in 1933.

Cover of the August, 1965 issue of *CQ Magazine* which ran the featured story.

Copyright and reprinted with permission of CQ Communications, Inc.

A tale of an early sale

Reprinted by permission, *CQ Magazine*, August, 1965

By Ed Marriner

A Sunday afternoon during the hot, humid Iowa summer is usually not conducive to much activity except relaxing or recreation. The often oppressive heat and sunshine place a mantle of stillness over the landscape, broken only by a gentle breeze.

Such was the case back in 1932 when Benton White and his wife were driving through the Iowa countryside. The heat and weariness from travel led them to call a halt and find a hotel room to cool off. Passing a newsstand, Benton noticed a *QST* magazine which he picked up, casually glancing through the pages. By chance his eyes fell on an ad for "Collins Transmitters" made in Cedar Rapids by Arthur A. Collins.

With nothing more important to do, Benton decided to locate the Collins factory even if it was a Sunday afternoon.

Returning to the hotel to get his wife, they drove around town with her as navigator until they arrived at the given address, a house on a quiet shady street. Despite her protests, "You must be mistaken, there is no factory here," Benton decided to ring the doorbell anyway and see for himself.

A lady answered the door and replied to his question, "Yes, this is the Collins factory. Just a minute and I will call Mr. Collins." Benton returned to the car to inform his wife that he would be inside for awhile.

As he again approached the house he overheard a man talking on the telephone to a Mr. Miller and urging him to hurry over, saying "I think we have a customer." Benton had no intention of buying anything that Sunday afternoon, and as he silently debated whether to go through with his visit a jalopy pulled up outside with a squeal of brakes. Before he could change his mind, Benton was whisked inside and down into the basement; the Collins factory.

In one corner stood the furnace, a cat curled up next to the cool steel base. Around the walls were neat, clean workbenches. On one bench was a nearly completed transmitter, and as the three men gathered around it, Mr. Collins and his associate, the only other employee of the firm at the time, fired up soldering irons and started working.

Benton recalled that as members of the ham fraternity they sat on stools and chatted. He was impressed with the transmitter as a high quality piece of gear, which he decided then and there he wanted for his own.

Informed that the transmitter was being built for someone else, Benton nevertheless was able to negotiate an immediate deal when he produced on the spot the $97.50 which was the price.

Everything was apparently working out just as Mr. Collins and his partner had planned, and in a short time the three of them were loading the transmitter in Benton's car so he could take it home to Chattanooga, Tenn.

It is believed this was the first or second full-sized transmitter, not a kit, which the Collins company sold, and a forerunner of thousands of Collins electronic units used for many applications throughout the world.

Down through the years the transmitter remained a treasured possession of Benton White, complete in its original condition except for replacement of a condenser. He operated the transmitter for a good many years until it became semi-retired as a low power emergency and field day rig.

Several years ago Mr. White died. The story of this incident was related by Harry Heilbeck, and brought to the attention of Mr. Collins. With a brief lead, the Collins company started to track down the early production transmitter. It had changed hands since the death of Benton, but was located and arrangements were made to obtain it for an amateur equipment display room at the Collins plant in Cedar Rapids. ▪

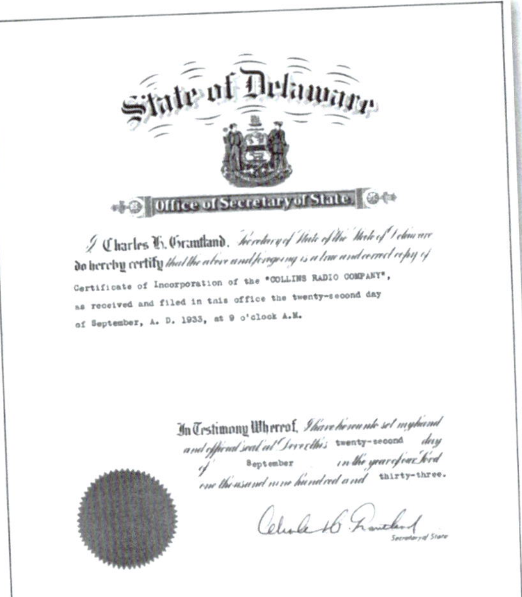

State of Delaware

Office of Secretary of State

I, Charles H. Grantland, Secretary of State of the State of Delaware do hereby certify that the above and foregoing is a true and correct copy of Certificate of Incorporation of the "COLLINS RADIO COMPANY", as received and filed in this office the twenty-second day of September, A. D. 1933, at 9 o'clock A.M.

In Testimony Whereof, I have hereunto set my hand and official seal at Dover this twenty-second day of September in the year of our Lord one thousand nine hundred and thirty-three.

Charles H. Grantland
Secretary of State

› The original document of incorporation listed the time as 9 a.m., September 22, 1933.

Corporation formed

On September 22, 1933, with eight employees and $29,000 in capital, Collins Radio Company became a corporation under the laws of the State of Delaware.

At that time Delaware had some of the most modern corporation laws in the country, and many businesses were officially organizing there, although their actual facilities were located in other states. (On May 13, 1937, the company reorganized as an Iowa corporation.)

The first meeting of the board of directors of Collins Radio was held at the new First Avenue location on October 10, 1933. The entire board of directors was present: Arthur Collins; his wife, Magaret; his father, M. H.; and his mother, Faith. Arthur was elected president, and M. H. was made vice president of the new corporation. Also present was Irene Snyder, who was elected secretary and treasurer.

Originally 250 shares of stock were issued for the company, which was given an arbitrary book value of $20,746. Of these shares, 124 each were issued to Margaret and Faith Collins and one share each to Arthur and M. H. Collins.

During the first two years, Collins Radio built only four types of transmitters, all designed for the amateur radio market.

But the reputation of the small firm grew, and Collins found that his equipment was being put to use by government and commercial enterprises as well.

However, outside the world of radio few people knew much about the new company, even in Cedar Rapids.

Jiggs Ozburn recalled that when Collins was called upon to install a new high fidelity system at Cedar Park, the dance bands that played there "made fun of our little (then new on the market) crystal microphone, saying it was just 'a crummy single button carbon mike.' At this time the city of Cedar Rapids was hardly aware of Collins Radio."

Another incident which pointed out Collins Radio Company's anonymity in the early days is told by long-time employee Millie Lahr.

"My brother, Albert Fay Rathbun, worked for a freight company. He was to deliver some parts to someone by the name of Arthur Collins in a basement and he was told to be sure to collect freight charges, because it was probably some fly-by-night outfit."

› Arthur Collins, president of Collins Radio Company, 1934.

› M. H. Collins, vice president of Collins Radio, 1934.

The big break

Late in 1933, a series of events began that would again put Arthur Collins in the national limelight, and assure his infant company a thriving childhood.

Most of Boston was still sleeping that October morning when a freighter of World War I vintage, the *Jacob Ruppert*, slipped from its moorings, inched cautiously through the harbor's mist and sailed southward into history.

Crammed into the radio shack of the ship was communication gear that would keep the 115-man expedition in contact with the rest of the world.

> In a room called the design laboratory, engineer Roy Olson worked at the table and Leonard Vick manned the drafting board. (1934)

The destination — the South Pole. The man in charge — Admiral Richard E. Byrd. The same Richard Byrd who learned about Arthur Collins from the Arctic in 1925 had become Rear Admiral Richard Byrd by 1933. The radio equipment — supplied by Collins Radio Company.

The black electronic boxes served as much more than a means of communication for Byrd. They played an unusual role as "fundraisers" for his second expedition to the Antarctic.

After CBS officials investigated the technical hazards of broadcasting from an area of the world well-known for the havoc it played with radio transmissions, they decided to risk underwriting the project, and commissioned the Collins Radio Company to design and build the transmitters and accessories.

Just eight years earlier the nation had read how a 15-year-old boy kept in daily contact with MacMillan's Arctic expedition. Five years later, in 1930, Collins again made communication news when his equipment provided continual contact with Admiral Byrd's first expedition to the Antarctic. For those two trips, the means of communication was shortwave radio telegraph. Because of the tremendous distances it was necessary to cover with relatively low-power equipment, the only radio was the more reliable dot-dash coded method of radio telegraphy.

But by 1933, advancement in shortwave radio voice communication (or radio telephony as it was called) was so great that Byrd decided to take radio telephony equipment along to the South Pole.

> Rear Admiral Richard E. Byrd.

The attempt was a gamble, for if it failed, the reputations of the expedition, CBS, and Collins Radio Company would suffer.

"It was a costly gamble for everyone if it failed," Byrd later wrote, "for involved was a 10,000-mile radio telephone circuit and expensive amplification relay set-ups — all from our ships and a shore station incapable of powering a transmitter with much better than a peanut-stand strength compared to the power in a large broadcasting station."

A careful study of the technical problems involved was made by Dr. T. S. McCaleb of Harvard University, director of communications for the expedition, and E. K. Cohan, technical director of CBS. So great was their faith in the plan and in the Collins equipment they had selected, CBS sold the proposed broadcasts to a prominent advertiser, General Foods Corp., so they would be heard not only by amateurs and shortwave listeners, but by the vast radio broadcast audience over the nationwide CBS network.

Major problems had to be overcome to make the plan work.

First was the limitation imposed by the source of power available. Consideration of fuel consumption for generators and weight limited the size of the transmitter which could be used. The commercial transoceanic transmitters of the day

For his Antarctic expeditions, Richard Byrd employed the latest devices available to explorers — shortwave radio and airplanes.

The workhorse of the 1933 expedition, a Curtiss-Wright Condor, was carried aboard the *Jacob Ruppert* and lowered to the water for takeoffs. The pontoons could easily be replaced with aluminum skis for snow and ice operation.

One member of the expedition prepares his gear for a trek on skis. The antenna in the background was one used for transmissions from Little America.

used 40 or 50 kilowatts output to maintain consistent communication. The study team found that the job would have to be done with a one-kilowatt transmitter.

Another problem was the antenna aboard the ship. Commercial ground stations used directional antennas covering acres of ground to concentrate as much of the signal as possible toward the receiving station. The inefficient antenna on the *Jacob Ruppert* would have to be used until the expedition reached Little America in the Antarctic.

The largest problem to overcome was the varying transmission conditions which affect shortwaves. Magnetic storms occasionally interrupted commercial circuits for days at a time.

But with these things firmly in mind, the engineers of the Byrd expedition went ahead with their plans.

Collins Radio Company shipped a large, one-kilowatt Type 20B transmitter to Boston for use as the main link with the United States. In addition, two smaller Collins 150B transmitters were used for communication between the different camps of the expedition.

Martin Kahn, Collins field engineer, assisted in installation onboard the *Jacob Ruppert* and accompanied the expedition from Boston to New York.

The first broadcast was made on November 11, 1933, from a position near Easter Island off the coast of Chile. The program was picked up in New York and placed on the network at the appointed hour. The CBS engineers at New York and the operators aboard the ship had been carrying out daily tests for two weeks, but this was the real test. As luck would have it, conditions were miserable. The *Jacob Ruppert* was in a heavy sea and the speakers had trouble keeping the proper spacing from the microphone because of the rolling of the ship. Nevertheless, the program came through understandably, although somewhat distorted by fading and background noise. Astonished broadcast listeners were thrilled when they heard the flagship's whistle, an introduction of various members of the expedition, the barking of the 150 huskies aboard the ship, and an announcement by Rear Admiral Byrd himself.

Later broadcasts were made from other positions in the South Pacific as the expedition journeyed on its way to Antarctica.

Postal Telegraph
THE INTERNATIONAL SYSTEM
Commercial Cables
All America Cables
Mackay Radio

C7 51 EXPED RADIO=
SS JACOB RUPPERT VIA MACKAY RADIO SAYVILLE NY 27 [illegible]
MR ARTHUR A COLLINS=
CEDARRAPIDS IOWA=
YOUR TWENTY B TRANSMITTER HAS BEEN OPERATING EXCELLENTLY FOR OUR BROADCASTS STOP AT OUR PRESENT LOCATION 6000 MILES FROM NEWYORK WITH GOOD ATMOSPHERIC CONDITIONS SIGNALS ARE RECEIVED WELL IN NEWYORK AND SANFRANCISCO STOP THE PERFORMANCE OF THE TRANSMITTER HAS LEFT LITTLE TO BE DESIRED=
R E BYRD.

> Collins received this telegram from Byrd as the expedition sailed toward the South Pole.

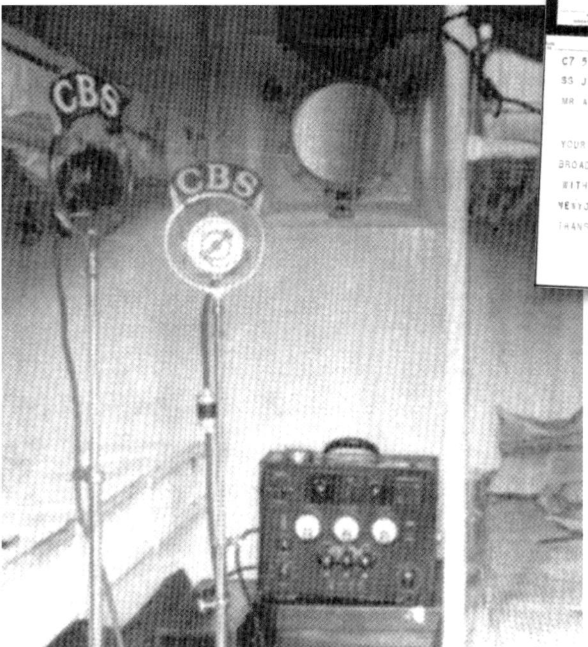

> Guy Hutcheson (left), radio operator on the *Jacob Ruppert* posed with one of the 150B transmitters used as an exciter to drive the larger 20B transmitter. The 150B also drove the antenna direct when full power was not needed.

> Top center: Broadcasts to the United States usually originated in other quarters, so crew members had room to gather around the microphones.

> Below: The large 20B transmitter used by the Byrd expedition was designed primarily as a broadcast transmitter, although some amateur operators who wanted maximum power also used it.

On February 3, 1934, Byrd prepared for the first formal broadcast from the Antarctic continent.

"We gathered in the Old Mess Hall by the dim light of kerosene lanterns. Hutcheson fiddled with controls, on a monitor board. He snapped his fingers across the microphones to test them. And I thought, as I watched these mysterious preparations, how broadly things had changed, how 22 years before, Scott and his whole party had silently died of hunger while his base, just 160 miles away, awaited his homecoming at Cape Evans, and here we were casually making ready to tell of our prosaic days to a vast audience in the United States. A thing called a cue came through. Each of us went to the microphone to say his piece and 10,000 miles away our voices came through clear as a bell. Somewhere in the shadow of the mess hall, 15 feet below the surface, it didn't seem possible."

Later, broadcasts were made on a weekly basis from Antarctica, and by July members of the expedition succeeded in establishing a radio link between the Arctic and the Antarctic using Collins equipment.

The gamble had paid off, for Byrd, for CBS, and for Collins. In as little time as it took an amateur radio operator to put out his call, word of the Collins equipment success spread through the ham world.

The Collins radios used by Byrd employed new ideas which were later widely adopted in the radio field. These included: multiple pretuned radio frequency bays which allowed the operator to make frequency changes more quickly than previously possible, and unitized construction which gave the radios top-notch quality and reliability.

Another feature of the radios was a system called Class B modulation, which allowed large audio power from relatively small tubes. Collins Radio Company virtually pioneered the application of Class B modulation in low- and medium-powered transmitters.

The performance of the Collins equipment in transmitting voices from Antarctica produced a boom in orders for equipment among both amateur and commercial buyers worldwide.

> Office group at the First Avenue building in 1934 included, left to right: Katherine Horsfall, Frances Power, Park Rinard, Irene Snyder, Faye (Jones) Dunn, Betty (Cocayne) Davis, Lucille (Eller) Dotson, and Marjorie (Black) James.

Growth in the 1930s

The publicity and prestige resulting from the Byrd expedition sent the company on its way toward expansion. A company newsletter, called *Collins Signal*, was started in January, 1934. The work force was increased.

One of those who applied for work with the growing company was Milo Soukup.

"Jobs were scarce," Milo recalls, " we couldn't even get into the Civilian Conservation Corps, which was the government jobs program. Leo, my older brother, got on at Collins because of the influence of his friend, Ted Saxon, and I also wanted a job."

Because he was only 16, Milo wasn't immediately given a factory job, but Arthur's father arranged for Milo to work around the family home on Fairview Drive, doing work around the house and driving M. H.'s Cadillac. A year later, Soukup was put on the assembly crew. His job was to sand and varnish wooden spars used in radio cabinets.

Other contracts came in for diverse applications of radio equipment. Unusual requests seemed to be the usual at Collins Radio Company.

There was Father Hubbard, the glacier priest, who had his Collins communications gear adapted for dogsled travel in Arctic regions.

There was the Maharajah of Mysore, who in 1935 ordered a Collins transmitter to take with him on a tiger hunt in India. He purchased four receivers — one for each wife — so he could talk to them while on safari.

Firestone Tire and Rubber Co. purchased an early transmitter for use in Liberia, Africa. There was no port at the designated place of delivery, so Collins employees sealed the packed transmitter in a large horse-watering tank and instructed the shippers to float it ashore. At high tide, a freighter carried the tank as close to the shore as it could, then tossed it overboard. At low tide natives towed it ashore with oxen. The radio replaced a former telegraph system because the natives had cut down the copper wire lines to create jewelry.

One of the best known customers in the ham community was Dr. James M. B. Hard of Mexico City. He chose Collins equipment in 1933 for his new installation, and had it specially designed for his needs. Dr. Hard spent weekends away from the city and wanted equipment he could take with him, so he commissioned Collins to design a special carrying case and distinctive equipment that would all easily fit into the trunk of his new Packard.

> Dr. James M. B. Hard with his portable transmitter, designed especially for travel.

Also in those early days radio equipment was produced for a Harvard expedition to Russia to study an eclipse, and for noted explorer Sir Hubert Wilkins in his search for missing Russian fliers in the Arctic.

To further advertise the capabilities of the growing company, a team of Collins "representatives" traveled to the 1933 Chicago World's Fair. The team of representatives actually amounted to the entire work force, except for the office women. "We thought our equipment was the hit of the show," recalled Jiggs Ozburn, although he admitted belly dancer sensation Little Egypt also stirred some excitement at the fair.

As world markets grew, several men were hired as Collins representatives to foreign countries. M. H. Collins and William Houk worked on contracts in Mexico in 1934, and the next year Raymond Billings and Frank Goodrich also made visits to Mexico.

› Cedar Rapids Police squad cars displayed in front of city hall, 1935. The Cedar Rapids Police Department was one of the first to use two-way radios in squad cars.

› The Collins 45A transmitter.

Jamie Lopez of Bogota, Columbia, was made the company's South American representative.

Ozburn recalled that Arthur would sometimes come out to his ham shack and talk business via radio with Clair Miller in Mexico City.

Two complete mobile communication stations mounted in two trucks were assembled for the Spanish Government during the Spanish civil war in 1937. Roger Pierce recalled that Spain sent two representatives who couldn't speak English to Cedar Rapids to supervise the work, and none of the Collins employees could speak Spanish, but the equipment seemed to work in either language, Pierce said.

Many police departments around the country were beginning to use two-way radios, and by 1938, Collins had supplied equipment for police in Lodi, California; Connersville, Indiana; Urbana, Illinois; Huron, South Dakota; and Ottumwa and Cedar Rapids, Iowa.

The radio most of those police departments purchased was a model Collins introduced in September, 1935, called the 45A transmitter. It provided increased power and improved efficiency in a compact table-top unit. By December, sales to amateur and commercial users were running high, and Collins increased production to meet the unexpected number of sales.

Several new Collins transmitters for broadcast radio stations were also introduced in 1935, although many in the industry were convinced the product wouldn't work well enough to sell. A Tennessee radio station was the first to install a Collins transmitter using Class B modulation, which was a radical departure from the modulation system that was then the broadcast industry standard. The equipment proved that adopting the Class B modulation theory not only required smaller tubes and other smaller components — which meant less space, less power and less expense — but that the end product, the transmission itself, was of higher quality.

With all the orders coming in, the company needed more factory and office space. In 1935, the factory portion was moved to a storeroom at the corner of Seventh Street and First Avenue. In 1937, the factory rejoined the office department at the 2920 First Avenue address, but by then the operation was too large for the original building, so the production force of more than 80 persons was moved to a building behind the offices.

› This 1935 photograph shows Henry Nemec, who later worked for Collins, using a 150B transmitter, similar to the one used by Admiral Byrd. Collins also built mobile equipment for police operations.

› A group of 100-watt lighthouse radiotelephone transmitters, LSR-319, built by Collins for the U.S. Department of Commerce, were lined up along the west wall of the building at Seventh Street and First Avenue, later occupied by Modern Laundry.

Total 1934 sales of $129,000 more than doubled in 1935, and increased to $391,000 by 1936.

With his growing reputation, Collins was able to attract highly skilled employees, many with extensive radio experience.

Robert Gates came to the company in 1935 and was made treasurer. Rose Hansen, the former secretary of the Collins Farms Company, became secretary.

L. E. Bessemer joined in 1935 and served as a principal development engineer. L. Morgan Craft brought with him a broad background in physics and electronics, and was a guiding factor in the company's design and development policies. Walter Wirkler collaborated with Arthur Collins on new modulation designs.

A former associate of Lee De Forest also came to work for the Collins Radio Company. William Barkley had become general manager of De Forest Radio Co. and Wireless Specialty Co., one of the earliest manufacturers of radio equipment. It was Barkley who coined the phrase "The Father of Radio" to describe De Forest. Barkley joined Collins in 1935 as executive vice president and sales manager of the new Collins sales office in New York, and later became executive vice president of the company.

One of the sales negotiated by Barkley that elevated Collins Radio to the ranks of the large communication equipment companies was a substantial contract with the Union of South Africa concluded just before the United States entered World War II. This sale was more than the total sales of any year in the company's earlier history. Later, this radio equipment was to have an important part in the battles of the North African campaign.

The year 1935 also brought another sign of the company's growth. Time clocks were installed.

Aided by a fellow scientist

During this pre-war period of growth, Collins Radio, along with other small and growing radio manufacturers, experienced difficulties in procuring components, especially vacuum tubes, from the major manufacturers.

Two of the largest companies, RCA and AT&T, made a significant move to squeeze out the small companies by claiming they had patent rights on vacuum tubes designed by De Forest and Dr. Edwin Howard Armstrong. RCA and AT&T had adopted the policy of not granting licenses for the manufacture of radio transmitters, and they sued Collins Radio Company.

Arthur Collins wasn't satisfied to sit back and watch his growing company be strangled, so he asked his patent attorney, John Brady of Washington, D.C., to make a search of the De Forest "file wrapper." After some research into the problem, Brady discovered that Robert Goddard, the famous rocket pioneer, had applied for and received a patent in 1915 on an oscillator tube that was an early form of the radio tube principle later employed by De Forest and Armstrong.

Collins appealed to Goddard to help Collins Radio beat RCA and AT&T in the legal action. In 1936, Arthur Collins traveled to Goddard's home at Roswell, New Mexico, where Goddard was engaged in rocket research and testing. The two scientists discussed the situation, and Goddard assured Collins he would consider his request for help.

While Goddard was considering Collins' appeal, RCA sent an emissary to Roswell to ask Goddard if his tube actually worked.

"Of course my tube works," Goddard replied testily to the RCA representative.

His inventiveness insulted, Goddard became ardent in his support of Collins. Armed with the help of the renowned Dr. Goddard, Collins was able to get RCA and AT&T to modify their licensing policies and grant a patent license to Collins, and to many other companies.

"Dr. Goddard didn't ask a thing for himself," Collins later recalled for Goddard's biographer. He accepted only a modest consultant's fee, saying it would be enough to pay for another patent.

Because of Goddard's assistance, Collins Radio Company was able to continue manufacturing and selling its radio equipment.

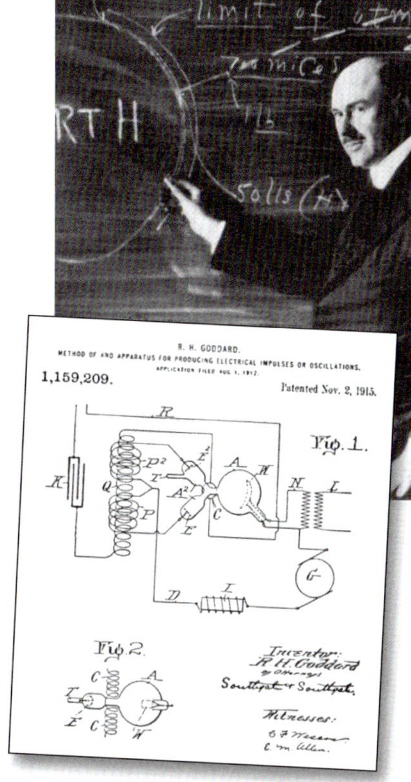

› Above: Robert H. Goddard. Photo from the Bettmann Archive, Inc. Below, Goddard's original vacuum tube patent.

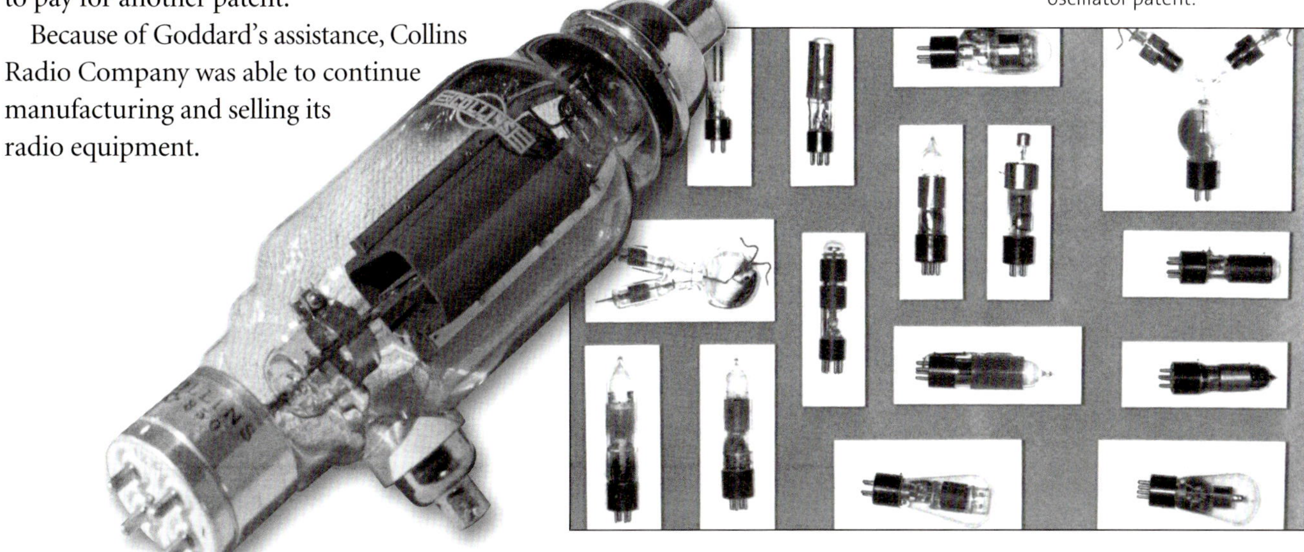

› This collection of tubes represented many of the types built by Collins in the 1930s. The "Y-shaped" tube in the top right corner was built as an exhibit in the RCA patent litigation as a copy of Goddard's patent, which was cited as prior art to the De Forest oscillator patent.

> A Collins transmitter was installed in the *Enterprise* in 1934, making it the first known airborne application of a Collins radio.

> The Goodyear blimp *Enterprise* as it appeared in 1935.

Something called the Autotune®

Growth of the company in the 1930s brought with it a broadening product line.

But from the beginning it was Collins policy to manufacture and sell only equipment of its own design and development. That policy was continued as the product lines expanded, leading Collins to become a major contributor to progress in the industry.

Collins Radio Company found that its radios, originally designed for the amateur radio market, were being applied in commercial and government situations as well.

Starting in 1934, Collins produced a 600-watt transmitter specifically designed for commercial radio. A line of broadcast transmitters also was introduced, starting in 1933 with a shortwave transmitter that was followed by a conventional 250-watt broadcast transmitter in 1934.

The first airborne application of a Collins radio came in 1934 when the Goodyear Tire Company's 12th commercial blimp, the Enterprise, was launched with a Collins transmitter on board. Large and heavy by today's standards, the transmitter provided effective communication for the airship crew stationed at Hoover Field in Washington, D.C.

Sales in the blimp market were sparse compared with the potential sales of radio for airplanes, which were quickly coming of age as commercial vehicles.

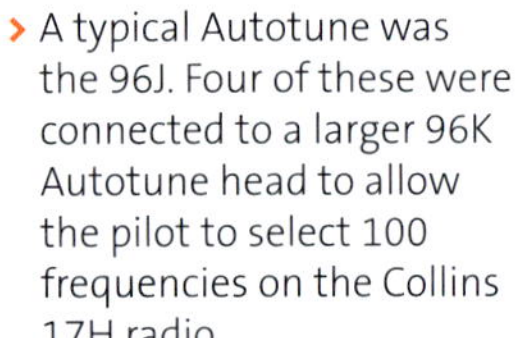

> A typical Autotune was the 96J. Four of these were connected to a larger 96K Autotune head to allow the pilot to select 100 frequencies on the Collins 17H radio.

A major problem airline operators faced in the early 1930s was the need to use a number of different transmitting frequencies within short periods of time. Different frequencies had to be used for plane-to-plane, plane-to-ground, and airport control tower communication. Changes in frequency were needed at night when changing atmospheric conditions made communications over a particular channel impossible.

But it took valuable time to re-tune a transmitter from one channel to another. A pilot coming into a field low on fuel couldn't waste time twiddling dials. Nor could he wait for a ground operator to switch frequencies.

How it works — The Autotune

To select a frequency using the Autotune, the radio operator simply dialed a frequency on the channel selecting switch (1). An electric circuit then caused the motor (2) to start turning the mechanism in the direction of the arrows, and by means of ratchets drove the seeking switch (3) and cam drum (4). The clutch (5) turned the positioned shaft (6) toward the home or reference end of its travel. The positioned shaft was directly connected to the tuning mechanism of whatever radio was being tuned. The shaft came to rest against the home stop. The clutch then slipped, allowing the motor to continue running. The limit switch (7) operated, allowing the motor to reverse as soon as the seeking switch found the selected position, and released the relay reversing the motor.

The cam drum and the seeking switch remained stationary. The controls were then set and the proper pawl (8) was in position. The shaft turned until the selected pawl engaged its corresponding stop ring (9) on the shaft. The positioning was then complete. The clutch slipped as the motor continued to run. The other limit switch operated and stopped the motor, ending the cycle.

For each preset frequency, a different pawl-stop ring-cam combination was used. They were located side by side on the same shafts, with spacers between them. Usually ten preset frequencies were possible. An airborne radio typically required five Autotunes driven by a common motor. Each Autotune was connected to a tuning device that required re-positioning for a frequency change. ▪

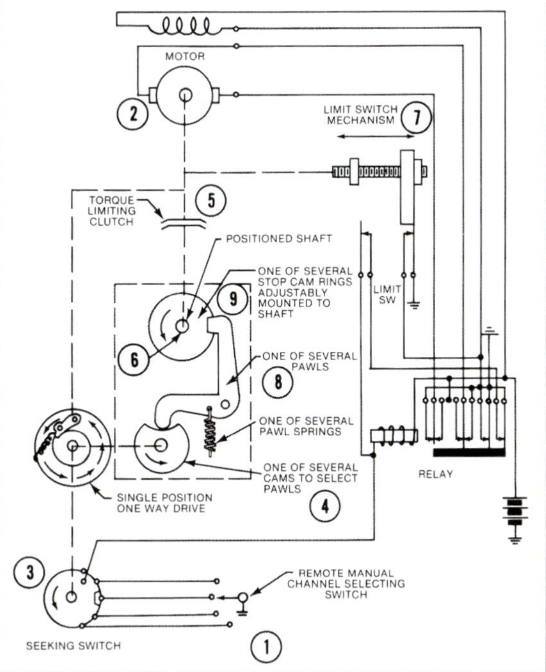

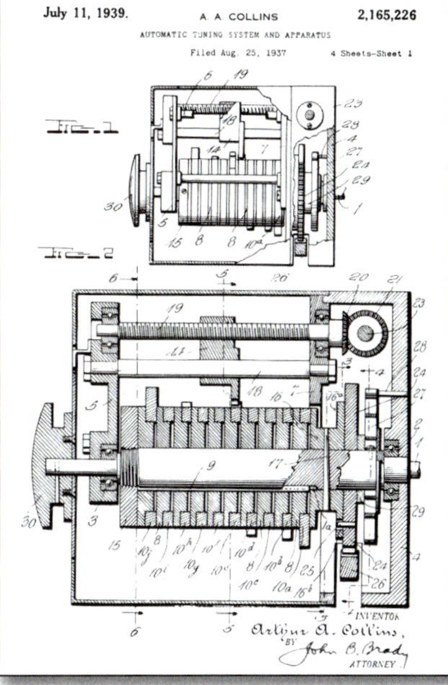

› The patent drawing for an advanced form of the Autotune showed the precision machining required to manufacture the device.

> Braniff Airways was the first airline to install Collins Autotune equipment for its fleet of DC-3s and other aircraft.

> The first airborne transmitter to use the Autotune was the Collins 17D, manufactured for Braniff.

The company and its 26-year-old boss tackled the problem, and hit upon an idea that was to have a major impact on the future of the company and an effect on the outcome of a world war.

For several years, Arthur Collins and his engineers studied the problem and worked on solutions for rapid radio band switching and tuning.

Then the solution hit.

Significantly enough, the solution occurred to Collins while he was on a commercial airline flight from Mexico City in 1935. Basically, the idea was a simple one — combine mechanical engineering principles with those of electronics. The theory was to link all of the tuning controls of a transmitter into one common dialing system. The Autotune® was born.

With the Autotune, switching from one band to another became as quick and easy as pushing a button.

The Autotune made it possible for pilots to easily control ten quick-shift channels at a time when airplane cockpits were becoming more and more complicated.

The 1937 edition of *Aero Digest* hailed the Autotune as "a major advance in aircraft radio design and construction," and the airlines were quick to see the benefits.

The United Fruit Company was the first customer, buying a six-circuit Autotune for its Central American ground stations.

Even though the Federal Radio Commission had a rule limiting aircraft radio transmitters to 50 watts, Collins developed a 100-watt transmitter which used the Autotune and sold it to Braniff Airways. Pink violation slips piled on Braniff's desk, but after a lengthy hassle, the government finally permitted Braniff and other carriers to raise their power. By 1937, Braniff became the first airline to equip its entire fleet with Collins Radio equipment. American Airlines also saw the importance of the invention, and became a major customer for Collins Autotune radios.

With the Autotune, Collins Radio Company entered the world of aviation. ▪

› (1) A 1938 publicity photograph taken at one of the First Avenue buildings showed 26 men working on the assembly line. In reality, employees were brought in from other work areas to make the assembly area appear larger than it was at that time, according to Milo Soukup. Those identified, outside row: Paul Hauser, Walker Whitmore, Kenny Vaughn, Harry Rogers, Dick Gintert, Buck Holsinger, Les Bessemer, Dale McCoy, unidentified, Archie Torson; Second row: Bob Davis, Ray Stoner, Soukup, Al Keyes; Third row: Tom McGregor, Bill Popek, Bruce Miller, Del Zarub, Chuck Hatfield, Claude Hoppe, Leonard Brawn, Elmer Koehn; Fourth row: Charlie Gould, Henry Dmitruk, Don Tubbesing, and Arlo Goodyear.

› (2) The production departments of the company returned to the 2920 First Avenue location in 1937. Instead of occupying the basement of the three-story building (top of photo) as they previously did, most factory operations took place in the larger one-story building immediately behind. This view looks east, with First Avenue in the background.

› (3) Robert Miller with a group of 7C amplifiers

› (4) Transmitter assembly area

› (5) Crystal holder department

› (6) Turret lathe

› (7) Spot welding

› (8) Cabling operation

› (9) Panel engraving

TCS YOU ARE ON YOUR WAY
TO DO YOUR DUTY FOR
THE U.S.A.

The war years

In 1939, the Nazi armies of Hitler invaded Poland. Despite the best attempts of American leaders to stay neutral in the European war, the Roosevelt administration was grimly aware that the United States was likely to be drawn into the battle, and that it needed to build military strength to effectively help its European friends.

The U. S. Navy was given the go-ahead to build a fleet second to none. Newly developed vessels — destroyers escorts, PCEs, flat-bottomed, shovel-nosed landing craft and baby carriers — slid down the ways and splashed into bays and launching areas. At the same time the Navy's air arm developed and expanded.

‹ The rugged little TCS was one of the radios mass-produced by Collins during World War II. More than 35,000 were built for use in small vessels and land vehicles. When the last TCS came off the production line, an informal ceremony marked the occasion.

Photo courtesy Kenneth Everhart.

› The Collins AN/ART-13 transmitter was credited with saving the lives of many allied pilots during World War II.

With a two-ocean navy and air arm underway, instant and reliable communication became a vital factor if this burgeoning fleet was to be welded into a trim fighting force. A system of communications that had been adequate for the needs of a small peacetime navy with experienced radio operators became clogged to the point of strangulation with the increase in operational traffic of a fleet preparing for war.

During this period Collins Radio Company, still less than ten years old, joined the Navy.

In the early spring of 1940, a board of naval air officers put prototype transmitters of three companies through a series of rigorous tests. Two of the transmitters were developed and introduced by companies long established and well-known in the field of electronics. The third transmitter — submitted by Collins Radio Company of Cedar Rapids, Iowa, "won the competition overwhelmingly," according to one of the officers conducting the tests. For the fleet air arm, the Collins AN/ART-13 offered a completely new field of flexibility. This was the first remote-controlled transmitter in naval aviation, and was the first time that radio equipment could be conveniently "pilot operated," permitting the remaining crew members to be at other action stations. This situation also saved enormous training load for the Navy, as one officer pointed out, since it did not require a radioman to be aloft on smaller aircraft.

> The Collins ART-13 transmitter flew aboard every B-29 Superfortress during World War II.

Boeing photo.

> The autotuned ART-13 was installed on fighter airplanes, such as the F4U Corsair.

Official U.S. Navy photograph.

> Many small U.S. Navy vessels such as PT boats were equipped with the Collins TCS radio during World War II.

Photo from the Bettmann Archive, Inc.

While other air-to-ground voice communication equipment of the period offered two- and three-frequency channels, the newly-developed Collins radio gave pilots nearly instantaneous selection of any one of ten frequencies.

A key design factor in this big step forward in naval air communications was the Collins Autotune, which permitted almost instant changing of frequencies.

The autotuned AN/ART-13 became the flagship of Collins Radio's efforts during the war. Collins produced 26,000 of the 100-watt airborne transmitters; and with other companies following Collins' designs and specifications, more than 90,000 were supplied not only to the U.S. Navy but to all of the nation's armed forces, plus those of the British.

A high ranking naval officer credited the ART-13 for literally saving the lives of American pilots in World War II. Just after Pearl Harbor, America's carrier-based planes were equipped with transmitters which had only four pre-tuned frequencies available. Each time an American Navy pilot took off in the first days of the war he was flying into a stacked deck, because Japanese intelligence was well aware of these four fixed frequencies. The enemy either jammed Navy transmissions or forced the carrier to keep radio silence on the four frequencies to avoid giving away its location.

Installed in Navy planes soon after the war began, Collins Autotune transmitters provided a turning point in naval flight operations. The transmitters offered any combination of ten automatically tuned frequencies. In 15 seconds a pilot could select

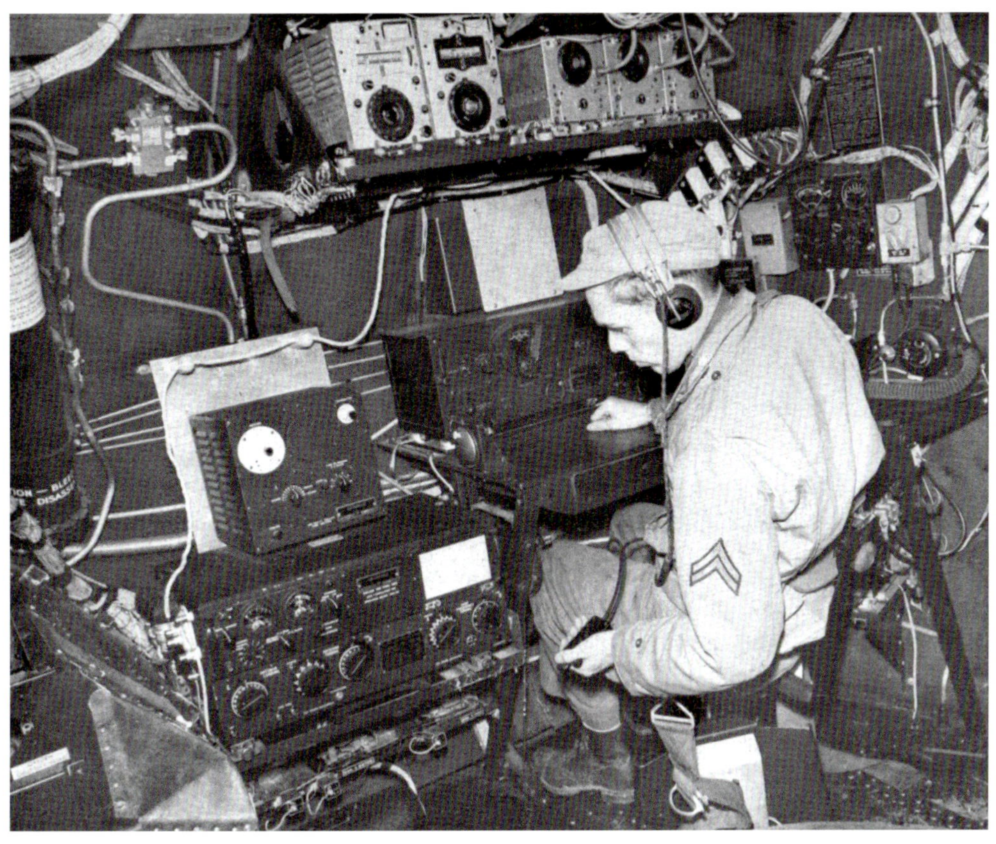

The ART-13 transmitter and antenna coupler were located on the radio operator's left in this view of a B-29 radio room.

any of the frequencies, and before each operation the ten frequencies could be changed, making it almost impossible for the enemy to monitor or jam American transmissions.

Developed during the same period was a compact combination transmitter-receiver designed for use aboard small vessels and land vehicles. Designated the TCS, this rugged little set was produced in quantities exceeding 35,000. It saw action in all theaters of the war — from PT boats skipping across waters of the Pacific to jeeps motoring along the autobahns of Germany.

Another Collins contribution during the war was a high frequency direction finder. A rotating beam-mounted loop antenna system, it played a key role in locating enemy submarines, particularly in the Atlantic. Collins tested this equipment in a two-story wooden blockhouse building nicknamed Fort Dearborn. At the time, the project was so secret that the doors of the assembly area were kept locked and admittance was granted to only a few employees who had been thoroughly investigated by the FBI. The direction finder was developed by engineers Gilbert Oberweiser, Willard Heath and Dale McCoy. Over a hundred of the units were made at Collins for the Navy.

Collins also built combination transmitters and receivers for mobile use during the war, and its large and powerful transmitters were used at Navy shore stations, air bases and army headquarters in all theaters of operation. It was via Collins transmitters aboard the U.S.S. Missouri in Tokyo Bay that the V-J Day surrender ceremonies were heard throughout the world.

A rotating direction finder was a top secret project at Collins during the war, and was used to locate German submarines.

Large five-kilowatt (cw) transmitters were built for the U.S. Navy. The radios, called the TDO and TDH, are shown being tested in May, 1944.

The little radio with a big task

A small Collins-built radio transmitter, buried deep in one of the tunnels of Fort Mills on Corregidor Island in Manila Bay, was the United States' only contact with its own and allied forces in the Pacific during the tragic early days of the war with Japan.

Several months before the Japanese attack on Pearl Harbor, the Commander-in-Chief of the Asiatic Fleet, Admiral Thomas C. Hart, directed the installation of an emergency Navy radio station in one of the tunnels of Fort Mills. This was in anticipation of probable destruction of all normal military communications facilities in the very early stages of the war, foresight that later proved to be correct in every detail.

The very first day of the battle for Manila saw all normal communication installations of the Army and Navy completely wiped out. Space, power and antenna limitations imposed by the concrete-lined tunnels on Corregidor restricted emergency station transmitters to relatively small and low-powered apparatus, including one shiny new Collins radio called a TCC.

› The Collins TCC radio.

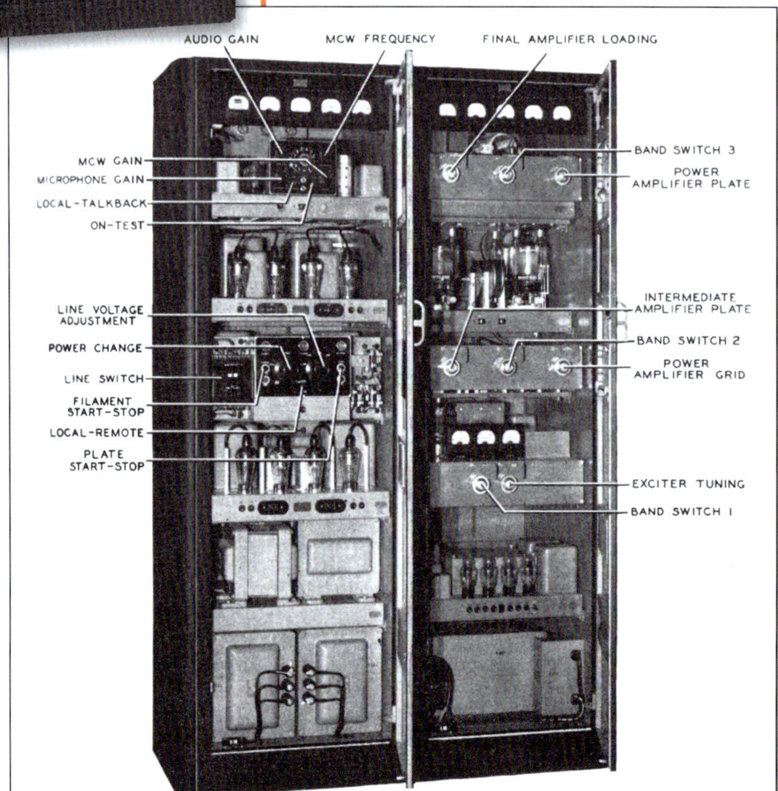

As General Douglas MacArthur made his famous "I shall return" statement before evacuating the island, allied forces retreated into the tunnels of Fort Mills in a final attempt to ward off the advancing Japanese forces.

After considerable experimenting, the remaining radiomen rigged an antenna with feeders that ran more than 100 feet before leaving the tunnel and turning down the hillside.

"Any radioman, viewing our weird installation, would have been convinced that we would be fortunate to get a signal across Manila Bay. Such, however, was not the case. The antenna system worked perfectly, even during bombing and bombardments when the antenna would be down or cut; our signal was not interrupted and we continued to put a strength five signal into Honolulu, many thousands of miles away." Lt. Comdr. C. A. Walruff later wrote:

"The most amazing part of the whole story is that the Collins TCC, which we all considered as very pretty in its nice enamel cabinet, very fancy with its telephone dial frequency shifting system, but probably not very rugged, turned out to be the most successful of all the apparatus available in the Far East. After our main radio station at Cavite was destroyed, the tunnel station on Corregidor took over all circuits, and the TCC was found to be the only transmitter which would work Honolulu. From that day on, until many hours *after* Corregidor had actually surrendered, every word that reached the United States from China, the Philippines, Singapore, Java and from our Asiatic Fleet was poured into Honolulu by the little Collins TCC (one-kilowatt) transmitter, keyed by high speed automatic equipment.

"The Collins transmitter maintained overseas communications for the Commander-in-Chief of the American, British and Dutch Naval forces and for the Army, as well as occasionally participating in the broadcast schedules for our submarine forces, which were even then causing tremendous losses for the Japs.

"It was not until after Java had fallen, and our forces had retreated to southwest Australia, that this tremendous responsibility was lifted from the brightly enameled frame of the very reliable, high-performance Collins TCC." ▪

Tremendous growth

When Collins Radio Company received the first orders for military radios in 1940, total employment was approximately 150. Many of the production workers were ham radio operators who had been attracted by the reputation of Collins equipment.

Company leaders realized that new government orders meant Collins would have to undertake a large expansion program to meet the requirements of the contracts.

The company also took a hard look at its long-term prospects, realizing that the electronics industry was about to branch into many new fields. In 1941, chief engineer Morgan Craft began a selective search for top young engineers who could lay a foundation for future growth. Of the approximately 15 who were hired, many made significant contributions to Collins Radio, and later, other firms.

Included in the Collins "Class of '41" were William Anderson, Robert Cox, Melvin Doelz, Mike Fitzgerald, Richard May, John Nyquist, Ernie Pappenfus, Ross Pyle, John Sherwood and others.

On November 4, 1940, the articles of incorporation were amended to increase the authorized capital stock from 250 shares to 1,000 shares of common stock. To provide funds to undertake expansion plans, the company took a mortgage of $100,000 on its real and personal property.

In the fall of 1940, construction of a new 52,000-square-foot plant started on a 26-acre tract between 32nd and 35th Streets N.E. in Cedar Rapids. The area was formerly a swampy pasture on the edge of town. Residential neighborhoods extended northward to about 29th Street at the time of the initial construction.

By January, 1941, many of the company's departments had moved out of the facilities at 2920 First Avenue and into the new 35th Street plant. Six months later a second construction project — an addition to the plant on the south and east sides which would more than double the size of the building — was announced.

> Main Plant, January 1941.

> Floor space of the Main Plant was doubled during the fall of 1941 when additions were built on the south and east sides. This view faces northwest.

> Model shop operators were fully equipped for all types of precision work.

> Engineering lab at the Third Street building, 1945.

> Some lab operations, such as receiver testing, required screened-in rooms to eliminate interference. August, 1942.

The $500,000 structure was built under an emergency plant facilities contract, with the U.S. government to reimburse Collins for the cost of the addition.

The last pieces of furniture for the new engineering areas and machinery for the factory area were being moved into place as word of the Japanese attack on Pearl Harbor reached the mainland.

The growth of Collins Radio suddenly began to mushroom. There was a war on, and the patriotic call went out across the country. Those who couldn't serve in the armed forces were expected to contribute at home, and for many that meant work in a war plant. Because of the radios built for all branches of the military, Collins became the largest war plant in Cedar Rapids. Amateur and commercial product lines were shut down, and all efforts were channeled into military equipment. Men who never intended to be supervisors found themselves heading large groups of workers. They had to learn by trial and error because there were no management courses in the early months of the war.

On December 22, 1941, the production requirements forced the company to cancel all leaves of absence during the holiday season. Operation on Christmas Day was "restricted" but not shut down completely for both office and factory workers.

More workers were hired in attempts to keep up with the increasing production quotas of the military, and vacations were cancelled.

Out of necessity, the company developed written procedures to assure uniformity for all aspects of production. There simply weren't enough instructors to go around to train new people by word-of-mouth, as had been the pre-war practice.

In February, 1942, the first women production workers were hired at Collins. A memo dated January 22, 1942 attempted to assure the male workers: "There will be no replacement of present male employees by women." Women of all ages helped fabricate and assemble radio equipment.

At first the women were stationed in a special area separate from the men — "penned up," as one employee recalls, "so they wouldn't be harassed by those nasty men." When management felt women and men could work side by side, restrictions on working areas for women were lifted. By the time the war production reached its peak in 1944, more than half the Collins assembly workers were women. To encourage more to apply, a short shift practice was adopted so housewives could work from 6:30 p.m. to 10:30 p.m.

Two women were hired as matrons for the women factory workers. Isabell Duncan and Leulla Chapman assisted the women at the plant and handed out smocks for them to wear over their clothes.

Milo Soukup, a foreman during the war, recalled one morning when five girls from the Amana Colonies (a disbanded German ethnic society near Cedar Rapids) reported for work in his department. "I never saw five girls that worked as hard. We always asked the personnel office if they had any more from the Amanas.

"It wasn't unusual for 18 new employees to be brought into a department twice a week," said Soukup. "We were told to put them to work. We were glad to have them, because other industries were also competing for people. We seemed to get good quality workers because they were screened by the personnel department."

› ART-13 drill press section, April 1944. The health message on the poster above the woman at left states "Coughs and sneezes spread diseases, please the Germans, and Japaneses!"

In January, 1942, Madge Taylor (Nabholtz) and Ruby MacArthur (Pyle) were hired as the first registered nurses at the company. A third nurse, Clela Dugan, was hired later that year.

Each day more workers were brought in from the surrounding area and put to work, some commuting as far as 60 miles one way. In the early days of tire and gasoline rationing, the personnel office set up a transportation department to make sure war plant employees received their allocations to get to work. The company's transportation director, Lucille Horner, arranged car pool rides for 1,644 persons in 393 vehicles. The average number of passengers per car was 4.2, the highest occupancy rate in the state.

The first copies of a new company publication, *Collins Column* were mailed in May, 1943. The newsletter, edited by Marian Kimball, was designed to be a morale booster for employees serving in the military.

A new cafeteria was constructed south of the 35th Street building, which by then had come to be called the Main Plant. Opened on September 6, 1944, the cafeteria had seating capacity for 488 persons. An outside company managed the cafeteria.

Collins also worked closely with Cedar Rapids City Lines bus company to schedule routes to the many Collins Radio plants.

The physical separation of the departments made an elaborate telephone system necessary. Three complete exchanges were installed, each large enough to serve an average-size Iowa town.

Despite the hardships of rationing and long work days, morale during the war was excellent.

› The Collins Radio accounting department December 31, 1943.

› Top: Assembly area for the TCS radio. This view was taken from the north end of Main Plant in October, 1944.

› Middle: The new employee cafeteria near the Main Plant. September 1944.

› Right: Interior of the new cafeteria. According to the menu posted on the wall, the plate lunch of pot roast and noodles, mashed potatoes or beets, plus bread and butter cost a total of 40 cents.

> Collins employees who shared the ride to work during the war had the highest per-vehicle occupancy rate in the state of Iowa. One of the more crowded vehicles was a panel truck driven by Tom Holland of Independence. Pictured from left: Holland; Marjorie Holloway, Ruth Willer, Floyd Holloway, Ira Wheeler, Keith Kennedy, Homer Kennedy, Leo Rogers, Don Holloway and Howard Weigand, all of Center Point; LaMar Guernsey of Urbana; and Sherman Hovey, Harley Dingsley, Walter Scott and Guy Beatty of Independence.

"The workers all felt they were part of the war effort," Soukup said. "There was no dissention, no complaints. It was a time of nice cooperative effort. It was all our job, not the other guy's problem."

Soukup pointed out, however, that the average person on the line during World War II probably didn't understand everything about the products he or she was building.

"One person was asked by a plant visitor what she was working on. She replied that she couldn't tell because it was a 'military secret.' Actually it wasn't a military secret, but our attitude was that we had a job to do and we did it. What the product did, well, that was up to the engineers."

The dedication to the war effort was recognized by the military when Collins Radio Company was awarded the coveted Army-Navy "E" emblem. On September 19, 1942, 4,000 persons gathered on the front lawn of Main Plant to watch William Willson and Alice Rinderknecht accept the pennant on behalf of Collins Radio employees from Brig. Gen. Charles Grahl, state director for selective services. Standing in a special reserved section facing the platform were family members and the 1,200 employees of Collins Radio.

> Army-Navy "E" emblem, circa 1942.

"Our job is to provide our full share of the lightning-fast radio couriers which will carry vital dispatches, commands, and reports to and from the growing thousands of ships and airplanes fighting a tremendous war extending over the entire world," Arthur Collins said in a speech to the crowd. "To do this job each one of us must increase his skill, improve his workmanship, speed his hands, and sharpen his wits. We must not only do the work at hand, but also prepare ourselves for many new problems to come."

Because of continued improvements in production, the company's "E" emblem award was renewed five times.

Employment numbers grew to keep pace with rising quotas, until a wartime peak of 3,332 was reached on July 15, 1945. As the summer progressed, it became apparent that the Allies would be victorious. Victory in Europe was followed by victory over Japan, and Collins employees celebrated.

When all was tallied, wartime contracts at Collins totaled $110 million, including $46.5 million in fiscal year 1943-44. But because of tight government pricing controls, the profit margin for the work was relatively small — a scant 0.4 percent.

> Phil Evans and John Grahm in the chemical laboratory. The lab, a group of test personnel, and assembly lines were located in the basement of the Shrine Temple. April 1944.

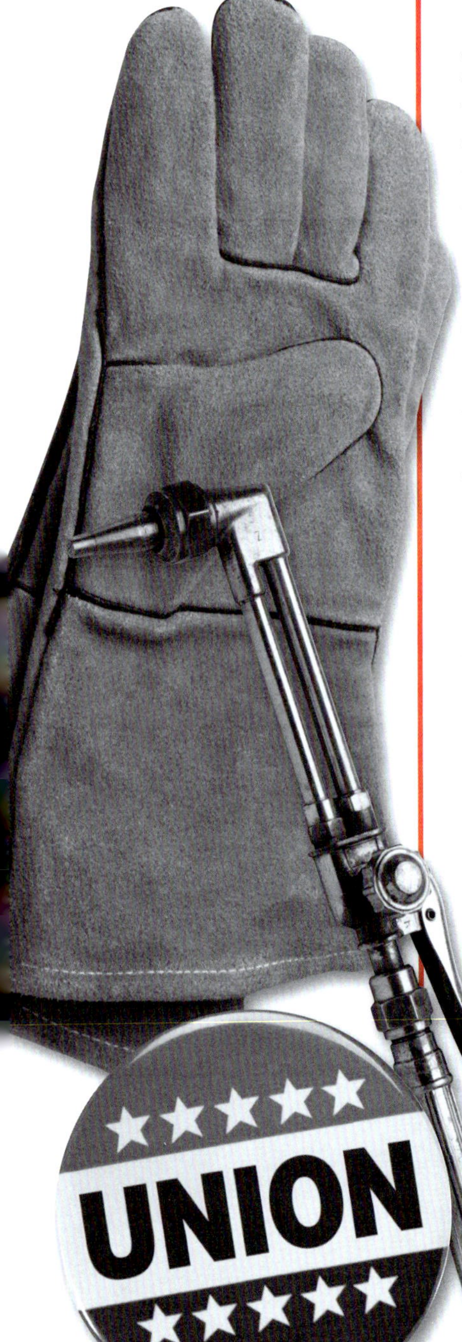

Workers' unions

During the height of wartime production at Collins Radio Company, production lines swelled with new employees. In line with the growing national labor movement, union organizing began at the plant.

In February, 1943, the company signed an agreement with an organization called the Collins Radio Employees Association (CREA). The contract included provisions for treatment of grievances, working conditions, negotiations, seniority rights and privileges, vacation policy and a bracket pay system.

But in May of that year the American Federation of Labor (AFL) and the Congress of Industrial Organizations (CIO) filed separate complaints, charging that the CREA was not a legal collective bargaining unit.

The National Labor Relations Board agreed. The NLRB ruled that because of the rapid expansion of the company during the previous two years, many organizers of the CREA had been made supervisors at the time of the contract. Therefore the CREA was ruled to be a "company-dominated" union, and NLRB recognition was withdrawn.

An election was held August 3, 1943, for production and maintenance employees to choose a new collective bargaining unit.

The results: International Brotherhood of Electrical Workers (affiliated with the AFL), 1,095; Radio Union, 717; Collins Radio Union, 57; and "no union" votes, 92.

The first officers elected for Local 1362 of the I.B.E.W. were, Bernard Rankin, president; John Lyons, vice president; Abe Mashman, first inspector; Lois Wallace, recording secretary; Edwin Pitts, treasurer; Frank Ross, financial secretary; Joe Buresch, foreman; and John Mitch, second inspector.

The first contract was negotiated October 11, 1943. The agreement raised the base wage to 57 cents and the maximum to $1.65 per hour, from the previous marks of 45 cents and $1.20, respectively. Equal pay was established for men and women doing the same job.

A steward system was provided to handle grievances, and holiday and vacation benefits were included in the contract.

In 1947, the AFL and the CIO joined to form a new national organization, the AFL-CIO.

A pension plan for bargaining units employees was negotiated in 1949.

Local 1362 grew from 68 members to 448 by 1948, and reached a peak of 6,500 in 1967.

In October, 1967, a labor dispute, primarily over wages, caused a strike by Local 1362 which lasted 16 days. In May, 1976, another strike, principally over cost-of-living increases, lasted 29 days.

Past Local 1362 presidents: Bernie Rankin, Abe Mashman, Joe Frycek, Wayne Disterhoff, Louis Schlatterback, W. D. Neff, Toby Arnold, Jerry Beer, Pete Jurgim, Craig Hoepner and Pat Marshall (present). Past business managers: John Lyon, Paul Anderson, Al Meier, John Hunter, Doug Heiden, Gary Heald, and Norm Sterzenbach (present).

Two other unions also represent workers at the Collins Division in Cedar Rapids.

Teamsters Local No. 238, a general local union with about 4,000 members, first represented employees in the data processing group in 1964. At that time there were 164 employees in the unit. Some of the early members were Vern Johnson, Ella Holland, Gene Poppe, George Abodeely, Jobie Atwater, and Carl Garrels. During the late 1960s Collins tape librarians and distribution clerks voted to become part of the unit. In 1974, the data conversion operators also joined.

The charter organizing a security guard union was signed January 12, 1945. The International Brotherhood of Electrical Workers Local 1429 elected Paul Walshire as its first president. Succeeding presidents were: Marvin Kuba, Harry B. Hall, Everett Freeman, Don Crowley, Robert Schoon, Al Herman, and Carolyn Duffy.

> To celebrate meeting the December, 1943 quota in the ART-13 assembly area, employees decorated the 1,500th radio with ribbons. Elmer Koehn, superintendent of the department, wore a ribbon on his head for the photograph. Standing behind the "1500" sign is Arthur Collins, and to Collins' left is J. W. "Slim" Dayhoff, production superintendent.

> Packed transmitters awaiting shipment to the Navy, May 1944.

The dollar figures only partially indicate the significance of the company's wartime achievement. Since all the products were of its own design and development, Collins assisted several other companies in the production of Collins-designed military equipment during late stages of the war.

There was also the ultimate human contribution made by four Collins employees in the armed forces who were killed in action during the war.

- Pvt. Denver Baxter: October 9, 1943.
- Capt. Patrick Casey: March 5, 1944.
- Lt. Melvin Forey: June 8, 1944.
- Cpl. Leonard Modracek: June 14, 1944.

The men's names were given special positions on a large plaque, which hung at the Main Plant, listing all Collins employees who served in the armed forces during the war.

Another loss felt by the company was the death of M. H. Collins, vice president, in April, 1943, at the age of 64. ▪

> An honor roll, which listed the names of Collins employees who served in the armed forces during the war, was placed near the east entrance at the Main Plant. Standing next to the plaque is Tim O'Brien, chief fire guard for Collins.

Leased buildings

By 1943, Collins had leased nearly every available building in Cedar Rapids. The government required many of the businesses which owned the buildings to cut back activities if they were not involved in war production. At one point, Collins had departments in 23 leased buildings.

Some of the buildings had unique characteristics which had a lasting impression on the people who worked there. The large Shrine Temple, for example, housed the Autotune assembly in its basement and a roller skating rink on the first floor. Mary Henningson said she clearly remembers how the music of the roller rink above the assembly area filled the building as the night shift worked to meet quotas. ▪

› (1) The Shrine Temple, at the corner of A Avenue and 6th Street N.E., was wartime home for the chemistry lab, test equipment design and assembly, and Autotune assembly. The building was also used as a National Guard armory.

› (2) Assembly area in the Shrine Temple. April 1944.

> (3) The Midland building was the location of sheet metal and finishing departments for Collins Radio.

> (4) Calder's Van and Storage building on A Avenue was used as a warehouse.

> (5) The Smulekoff building housed Collins' spare parts department.

> (6) The Chandler Company buildings were used as warehouses.

> (7) Receiving and inspection departments were housed in the Wagner building.

COLLINS RADIO COMPAN

Post-war restructuring

The end of the war brought drastic changes to Collins Radio Company.

During the previous four years the company had only one customer — Uncle Sam. With the war over, military contracts were cancelled or reduced. Of the original $47 million in orders on the company's books on July 31, 1945, $15 million remained after contract terminations and reductions. The remaining backlog represented advanced design programs requested by the military.

‹ The Collins hangar with a Twin Beech 18 in 1946.

› The message rang loud and clear on radios and televisions across the globe! News of the German surrender broke in the West on May 8, 1945, and celebrations erupted throughout Europe. In the United States, Americans awoke to the news and declared May 8 V-E Day.

War contracts were terminated, and that meant physical conversion of the company's plant facilities for peacetime operation, accompanied by large-scale cutbacks in the work force. The company needed time to adjust to manufacture new product lines, and employees had to retrain.

Among the other war plants in Cedar Rapids, not all were as critically affected by the government contract cancellations. An August 17, 1945, *Cedar Rapids Gazette* story summarized the condition of local industries. At Century Engineering, two-thirds of the government contracts were cancelled. Cherry-Burrell quickly converted to peacetime production, and advertised for machinists and sheet metal workers. Quaker Oats continued to supply military provisions, but was anxious to resume its civilian business. At the Wilson and Company packing plant, no drop in employment occurred because the demand for meat increased after the war. Demand for highway construction equipment kept Iowa Manufacturing and Iowa Steel employment levels high. Similar stories were told for Link-Belt Speeder, Turner Company, National Oats, Penick and Ford, Freuhauf Trailer, and LaPlant-Choate. Employment levels at most of Cedar Rapids' factories remained high, primarily due to the nature of their businesses, and also because most had been able to continue producing and selling commercial goods during the war.

At Collins, however, the story was different. Production had been completely dedicated to the armed forces.

In August, 1945, Collins Radio announced a temporary five-day layoff of all production departments as management struggled to determine future employment needs. All leases for space no longer necessary for operations were terminated.

> A group of veterans who returned to Collins posed beside the Main Plant.

One week later Robert Gates, company vice president, announced that within 30 days 2,000 employees would return to work, with the remaining 1,000 to be laid off for an extended length of time. Most of those laid off were women production line workers.

A survey made one year later showed that 79 percent of the women who had worked on the production lines at Collins had either quit voluntarily to resume pre-war lifestyles, or were laid off.

In June,1945, a testing program was initiated for personnel. Prior to that the method used to fill a vacancy or new position was to select the person with seniority considered to be qualified for the job. This resulted in a certain percentage of persons (an estimated one-third) being placed in positions for which they were not qualified. Three types of tests were used to measure an employee's qualifications: 1. aptitude tests, 2. achievement or performance tests, and 3. dexterity tests. The tests were established in cooperation with the union, using input from the officers and stewards.

The biggest single factor affecting the profit picture at Collins was the cost of developing and tooling new products. Because the company had no commercial products ready for sale at the end of the war, the development and retooling costs were exceedingly high in relation to income. The years 1946 through 1948 showed net losses in earnings for the company.

The huge retooling effort undertaken by Collins Radio was also reflected in employment and facilities. The Los Angeles sales office, established in 1946, was expanded and moved to Burbank, California, in 1949. Research and limited production facilities were added, and the Burbank facility became the Western Division of Collins Radio Company.

Roy Olson, Frank Davis and Arthur Collins check plans for a new aviation radio.

Priority job

On the morning of November 15, 1945, Arthur Collins, Frank Davis and Roy Olson were in conference. Collins had returned from a meeting at the Beech Aircraft Corporation at Wichita, Kansas, and had brought back with him an idea plus some sketches.

In the sudden hush of business following World War II, his company needed a commercial product to generate sales. At the same time the aviation industry was expecting a boom in private flying and needed a new, compact voice transmitter.

A survey of the possibilities indicated that a compact, straight-forward design could be built with components that the company had on hand. Ten minutes later the project was assigned, carrying an "A.A.C." (Arthur A. Collins) priority in the model shop and production departments. Rough sketches were used in lieu of drawings, and chassis construction utilized paper templates, with the layout drawn on the paper. Parts were arranged on the chassis and holes drilled on the spot.

Ten days after its inception, the model and a remote control box were complete and tested. The 17E-2 was ready for installation in a twin-engine or large single-engine airplane.

During the next week, inquiries for the transmitter by those who saw the model indicated the immediate need for the 17E-2. The first production run was set at 200 units.

The technical ability, skill and experience demonstrated by the short turnaround time reinforced the slogan of the day, "If it can't be done, Collins will do it." ▪

New personnel at Collins included some of the leading scientists in the fields of broadcast radio, radar, radio navigation, microwave communications, and aviation.

Marketing was given increased emphasis during the restructuring phase. A new marketing division was organized to cover all potential markets for the present and new products. W. J. Barkley, executive vice president, was a major contributor to the reorganization plans, along with Morgan Craft and Robert Cox. Vice President Robert Gates was placed in charge of the procurement and marketing division. An export division was also formed in 1946, with Robert Parsons as manager.

› The company displayed its wares at the 1946 Iowa State Fair. At left is a TDO transmitter. Behind it is an oversized wooden model Autotune (on the table). Behind George Price are microwave antennas, a microwave receiver and transmitter, test equipment, and two display boards with microwave tubes and "plumbing."

To finance the development, the company in November,1944, augmented its working capital by issuing public stock for the first time. Collins Radio Company sold 140,000 shares of common stock and 20,000 shares of preferred stock. The remaining 170,116 outstanding shares of common stock were retained by Collins family members and other managers. The sales added approximately $3 million to the company coffers.

By 1949, the new development efforts were paying off. Collins was producing specialized electronic equipment and was shaping for a future larger and much more complicated than its past. Two markets in particular were to form the basis for the post-war growth of the company in the late 1940s and early 1950s: ultra-high frequency radios for the military; and commercial aviation electronics.

Military radio

During World War II the hundreds of channels required in joint operations of all types of ship, shore and aircraft units in military emergencies overcrowded the commonly-used frequencies. With commercial airlines already committed to the very high frequency (VHF) band, the Defense Department decided to vacate VHF after the war and move military short range communications into the practically unexplored ultra high frequency (UHF) band. Besides availability of channel space, the UHF region was free of static and largely free of man-made interference. The military could count on reliable UHF communication within the "line of sight" transmission range regardless of weather conditions.

The U.S. Navy, at the close of the war, sponsored development of a UHF airborne radio, and Western Electric Company developed an experimental model for the Navy to evaluate.

At the same time, the Army Air Corps was evaluating its own new UHF radio developed by the Bendix Corporation.

Collins Radio Company was also working on UHF development models in the late 1940s in an attempt to catch up with the advantageous positions held by Western Electric and Bendix with their respective branches of the armed forces. At stake were government contracts worth millions of dollars.

When the evaluation periods for the new UHF radios were completed, the Navy recommended that a Collins-designed UHF radio be chosen as the Navy standard, but the Army still favored its own development model.

In the final analysis, it was a new doctrine adopted by President Harry Truman which affected the outcome of the contract decisions. The experiences of World War II revealed weaknesses because of competition and lack of coordination between branches of the military. Truman proposed the creation of a separate air force and a single cabinet officer in charge of defense. The idea was opposed by the Army and Navy, but in June, 1947, a unified National Military Establishment was created. As part of that unification effort, the new Department of Defense required that a standard design would be used for the new UHF airborne radios for all military branches.

Sealed bids were taken for the initial order of 4,000 radios for the Navy, and Collins won the $7 million contract on the basis of both quality and cost. Because of the unification policy, the defense department also awarded the Air Force contract to Collins.

Linear-tuned circuits developed by Collins in 1944 had an extraordinary effect on communications equipment, and the new UHF radios were among the first to employ them. These circuits, coupled with the Autopositioner,® another Collins development related to the Autotune, made complete electrical remote tuning a reality. A radio operator at a remote station could select literally thousands of frequencies, all produced with a relatively small number of crystals. While the VHF radios of the day could tune 10 to 12 frequencies, the new UHF radios provided 1,750 channels. The ARC-27, as it was designated by the military, was a transceiver, which meant that it both transmitted and received on many of the same circuits.

The ARC-27 was developed by a group which included J. P. Giacoletto, M. R. Hubbard, and Horst Schweighofer. Most of the detailed mechanical and electrical designs were done by E. K. Vick, Fred Holm, John Goetz, Emil Martin, H. Lehman, and Gordon Nicholson.

The Collins engineers originally planned a production rate for the Navy of 500 radios per month. However, with the Air Force contract award also going to Collins Radio, the quota was set at 1,000 sets per month.

Production began in late 1950, with the first 15 radios delivered to the military in January, 1951. By June, 1951, production reached 1,000 per month and continued at that level through September, 1952.

The success of the radio led to another military contract award to Collins — for UHF ground station radios to link with the airborne units. The Department of Defense had originally contracted a fourth manufacturer to build the ground station sets, but because of complications, Collins was asked to supply a complete redesign. Approximately 15,000 of the ground station

› AN/ARC-27

> At the end of the production line for the ARC-27.

UHF radios, called the GRC-27, were produced using the same design concepts as in the highly successful airborne units.

During the same period, the Korean conflict escalated, United States military support was sent to aid the south Korean government, and pressure was increased to build more ARC-27 radios. To meet the increased quotas, Western Electric was subcontracted to produce 10,000 radios under the Collins design. Another radio manufacturer, Admiral, also joined in the huge production effort and produced 25,000 Collins-designed UHF radios. In Cedar Rapids, Collins Radio assembly lines turned out a total of 40,000 ARC-27s. The radio was installed aboard almost every U.S. military aircraft operating in the 1950s.

The demands of rapid production made the ARC-27 a pressure program for Collins. Full production was at times delayed by parts shortages, caused by the nationwide industrial scramble for defense materials, by problems of personnel procurement and training, and by production and testing difficulties involved in building a new and highly complicated product.

"From a production standpoint, the difficulties which limited the accelerated production quotas boil down to two major ones," said John Dayhoff, production superintendent in a 1951 *Collins Column* article. "These were the training of inexperienced personnel and the shortage of parts."

> Installation of a Collins ARC-27 radio in an F-86 fighter jet.

By June, 1951, the problems were overcome and the increased production quotas were being met. In recognition of the achievement, a celebration was held with speeches by Arthur Collins and Commander R. J. Wayland, Naval inspection service, and the coronation of an ARC-27 assembly queen, Kathleen Newkirk. Glenn Johnson, director of industrial relations, served as master of ceremonies.

In his speech, Arthur Collins praised his employees for meeting the quotas, and took the opportunity to answer some of his critics in the industry.

"Going back to 1945, I might remind you that our company at that time redoubled its efforts in the development of special military equipment, including work which led to the ARC-27. We stressed special equipment development throughout the years when other companies in the industry were busy with conversion to civilian production and were happily building television sets and other home gadgets. Many people thought we were crazy to continue our old line of work. As a matter of fact, we did have several unprofitable years of operation, but we were successful in developing several very important equipments, probably more than any other electronics company."

The impact of the UHF radio contracts on the company was tremendous. In 1948, total sales were less than $8 million. In 1951, ARC-27 production totaled approximately $12 million, and in 1952 the figure increased to $24 million.

› A genius at work — Arthur Collins was referred to by many as a genius in the world of radio communications. The founder of Collins Radio Company was known to stay at work for 24 to 36 consecutive hours. During those times, he would take "catnaps" on his "scrupulously spotless desk."

Aviation Electronics

The second major post-war growth market for Collins Radio Company was in the field of civil aviation.

Before the war, a few airlines flourished with the widespread use of reliable and relatively comfortable airplanes such as the Douglas DC-2. Armed forces air shuttling during the war gave hundreds of thousands of persons their first taste of air travel, and there was no going back.

Another area of civil aviation development was business flying. American business sprouted wings at the close of World War II. Several reasons have been advanced for the post-war business flying boom. Air-mindedness was a primary cause, but it took a couple of catalysts to put companies in the flying business. One was the large number of surplus government planes — Lodestars, C47s (the military version of the DC-3), B-17s, B-25s, and others offered for sale at cut rates at the end of the war. The excess profits tax and favorable tax provisions for amortization made the heavy outlay for aircraft and air facilities less painful.

In the late 1940s and early 1950s, aircraft began operating in the improved communications and navigation environment brought on by advances in electronics technology. Air traffic control radar, instrument landing systems, and VHF communications aided in control and separation of aircraft operating in poor weather conditions. A new system of navigation called VOR (VHF omnidirectional range) was installed in ground stations across the country to replace the limited coverage, low frequency "beam" navigation system. The world's skies became more crowded with aircraft flying in worse weather conditions than ever before, causing a great increase in pilot workload, nearly approaching human limits.

Simple, easy to operate, reliable radio aids held a large part of the answer to the increased safety of air travel. Such electronics aids had to provide good position, steering, and weather information, with a minimum of attention and operating skill on the part of the pilot.

Recognizing the needs and the potential of the growing markets of business and commercial aviation, Collins moved to expand its aviation research and development. The Collins philosophy at the start of the effort was based on two key points: 1. Make it easy for the pilot. 2. Make the equipment so reliable that the chances

› The Collins hangar at the Cedar Rapids Airport, May 1945.

of failure are extremely small. Collins engineers knew that only with this approach could the impulse to distrust "gadgets," mechanical or electronic, be overcome in the pilot's mind.

In 1945, a company hangar was built at the Cedar Rapids Municipal Airport, south of the city, to house two new company airplanes. The facility was also built for flight testing and proving advanced designs in radio equipment by actual installation and use in aircraft, and for service to customers. *(Note: A staff of about 40 Research Division scientists, engineers, and laboratory assistants were moved to the facility in late 1945 to work in the laboratories adjacent to the hangar. Their work dealt primarily with new radio propagation techniques. (See Chapter 5.)*

One of the first developments from Collins for the new thrust into civilian aviation was a 100-watt, 20-channel, high frequency communications radio called the 18S. More than 500 were delivered in 1947, making Collins Radio the leading supplier of airborne radio equipment to the airlines. Important early customers for the 18S included American, Pan American, Northwest Orient, Braniff, Colonial, and Peruvian Airlines.

Also significant in the emergence of Collins as the leading supplier of airborne radio gear for the airlines was a new navigation receiver called the 51R. In 1946, Aeronautical Radio, Inc., a non-profit, research and communications company jointly owned by the major airlines, asked the radio industry for proposals on airborne receivers designed to meet the difficult specifications demanded by the new VOR system of navigation. Collins was one of six companies to comply and one of two whose designs were approved. The receiver equipped commercial transport planes for the new VOR navigation system being installed around the nation. In addition to the navigation function, the 51R was used to receive the localizer signal of an all-weather landing system. The localizer transmission defined the centerline of runways equipped with the new Instrument Landing System (ILS).

› The Collins 18S high frequency radio.

Flying feats

Excerpted from the August 14, 1977 *Cedar Rapids Gazette*

By Art Hough

About two years after he was graduated from Iowa State in 1932, Walt Wirkler went to work for Collins Radio.

"At the time there were 22 people, including me and the scrubwoman and Arthur Collins' father," Wirkler said.

Asked to tell a yarn or two about his flights with Arthur Collins, whom he characterized as "an excellent pilot," Wirkler came up with this one:

"The company had bought a Beechcraft in Wichita, Kansas, and we were going down to get it, meeting a Braniff pilot who was to fly it up for us.

"We got Jim Wathan to fly us down in a Fairchild. We started out from Cedar Rapids in a helluva head wind. So we hugged the ground. It got so rough that Jim said:

" 'To hell with it. Let's take it up to 7,000 feet and buck the head wind. We'll have smooth air anyway.'

"He told me to take it up while he was looking up stuff on his maps. We'd just leveled off at 7,000 feet when the darned thing swallowed a valve.

"It was almost dark on the ground then, near Excelsior Springs, but at 7,000 feet we had time to look around. I finally figured we could land on a razorback ridge which was headed into the wind. But when I started heading it that way, Jim took another look.

" 'Supposing we got hurt and they couldn't get in there to get us out?' he said.

"So he took over and plopped us down in a little pasture with white-face cattle. They never even looked up, because we weren't making any noise. We had to get a farmer and his pickup truck and some rope and pull this thing into the corner and build a little fence around it.

"He took us to a road and Arthur and I hitchhiked to Excelsior Springs, got a bus to Kansas City, and an airline from there to Wichita. Jim got a ride with a hearse. He hitchhiked all the way home."

"You had to have a little sense of humor if you wanted to fly in those days. If you wanted to fly, you'd better not figure on getting to a certain place at a certain time. You might not get there at all."

Editor's note: Wirkler was prolific as an engineer and scientist for Collins between 1938 and the mid 1950s. He researched direction finder navigation, participated in early high-speed radio telegraphy studies, helped develop navigation computers, autopilots, instrument landing systems, and collaborated with Arthur Collins to develop the first horizontal situation indicator. He now lives in Garnavillo, Iowa.

Arthur Collins, Vernon Rittger and Clark Chandler picked up the company's second Beech Model 18 in Wichita, Kansas.

> To flight test omnidirectional range equipment, Collins purchased a large Douglas C-47 from war surplus. 1947 photo.

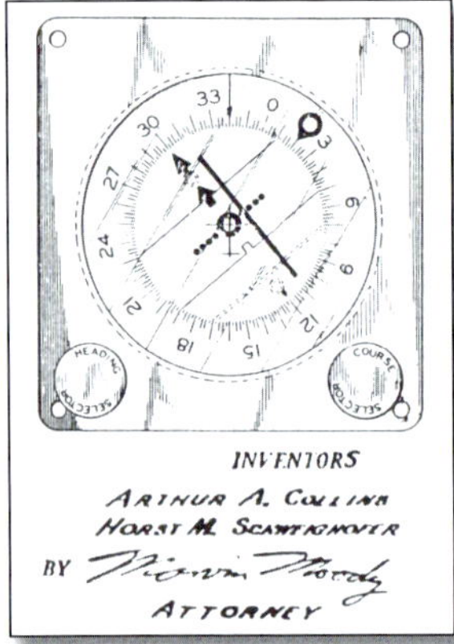

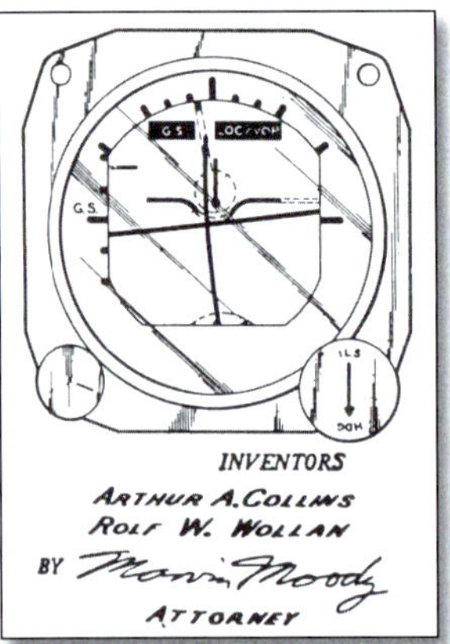

> The Collins integrated flight system consisted of the radio box, the course indicator and the approach horizon.

At this time, the company purchased from war surplus a Douglas C-47. The airplane was reconditioned and fitted with a model transport radio and pilot instrumentation set-up. By using the C-47 in flight tests with its equipment, Collins acquired knowledge of flight problems and used that experience to design effective airborne equipment, including the 51R.

One of the key Collins engineers who developed new aviation equipment was Francis L. Moseley. As a U.S. Army colonel during the war, Moseley developed the instrument approach system used by the military for blind landings, and was the principle designer of the Collins 51R.

To help the pilot cope with the increasingly complex problems of flying, Arthur Collins and his associates worked to develop a new flight director system, which was introduced in 1950. The Collins Integrated Flight System had several unique features provided by the instruments and a steering computer.

Radio navigation information was displayed on a course deviation bar, but unlike previous systems, this bar was superimposed on a compass card enabling the pilot to visualize directly his angular relationship to a selected course. This marked the birth of the horizontal situation indicator (HSI), which has been applied over the years in nearly every aircraft application, from private airplanes to jumbo jets to the Space Shuttle. The first HSI was patented by Arthur Collins and Horst Schweighofer, based on an idea by consultant Siegfried Knemeyer.

The HSI was paired with an attitude direction indicator which provided an artificial horizon with steering information superimposed. The new system replaced five cockpit instruments with two, and gave the pilot a clear pictorial presentation of the information he needed for en route navigation and precise instrument landings. The Collins flight director system freed the pilot from several mental computations, allowing him to do his essential job of flying more effectively. For example, the steering computer automatically corrected for crosswinds encountered during a landing approach.

Collins engineers and salesmen toured the country in the company's Twin Beech D-18 equipped with the new system and were met with enthusiastic receptions. The August, 1952, edition of *Flight* magazine quoted a chief pilot as calling it "a pilot's dream . . . the most complete and simplest instrumentation yet perfected to assist pilots in overcoming the problems of low approaches."

> The approach horizon instrument and a smaller version of the course indicator as they appeared in a cockpit. The integrated flight system instruments are second from the left.

> An early commercial airliner to have the Collins integrated flight system was the Vickers Viscount.

The first sale of the system was for a converted B-25 owned by Albert Trostel and Sons of Milwaukee, Wisconsin, and it was installed by Associated Radio Co., Collins Radio's distributor in Dallas, Texas. By 1952, Collins had received orders for 50 of the systems, which sold for about $4,000 each, and 1953 production was set at 1,000 systems.

The business aviation market was the first to embrace the new system, largely because owners of business aircraft were not encumbered by requirements to standardize equipment, and therefore could adopt new products more easily than could the commercial airlines.

By 1955, the airlines were also using the Collins flight director system. The first major airline application was aboard the Vickers Viscount airliner, chosen by Trans-Canada, Trans-Australian, Air France, Capital, and several others.

Using bearing information from two VOR stations, a navigation computer developed by Collins was a predecessor to today's area navigation systems.

In 1954, Collins introduced its first autopilot for automatic control in cross-country flight and during ILS approaches to airports. The AP-101 was designed to work with the flight director system so the pilot could continuously monitor the approach. The combination was an entirely new method of automatic flying which gave the pilot a continuous and easily understood picture during automatic control, yet allowed him to take over at any time and manually fly with the same flight director instruments.

Collins built several types of aircraft antennas to accommodate all of its airborne equipment. The most widely known antenna of the period was the "deerhorn" navigation antenna, so named because of its distinctive shape. Used with the 51R VOR receiver, more than 10,000 deerhorns were installed atop the cockpits of most non-jet commercial airliners, business, and military planes in the United States. The deerhorn was developed in 1948 by Collins engineer John Shanklin.

Collins introduced its first weather radar system in 1955. Delta Air Lines was first to install Collins WP-101 weather radar. Delta also chose two other new systems: automatic direction finding equipment, and Selcal, which permitted ground stations to call specific airplanes. The $1.5 million order from Delta came shortly after the airline was awarded new routes from Atlanta to New York, and from New Orleans to Houston.

Collins weather radar systems were designed to be installed in large airplanes, but a Dallas

> This cartoon in the December, 1951 *Collins Column* found humor in the shape of the "deerhorn" airborne antenna. It was drawn by Richard Pinney, now an artist in the Cedar Rapids area.

"It was a dandy shot, Swedley! I got him just as he lifted his wheels at La Guardia."

› CNI units were developed for new generation military jet aircraft.

› An early contract for Collins CNI equipment was for installation aboard the U. S. Navy's A3J Vigilante, built by North American Aviation.

television station, WFAA-TV, installed a system at its studio in 1958, and became the first to present local weather by radar to a television audience.

The transponder was another avionics unit introduced to aviation in the mid-1950s. As an aid for air traffic control, the Collins transponder provided a signal which reinforced ground radar to permit positive identification by the air traffic controller.

In the late 1950s, the new F-4 Phantom generation of military jets presented new problems for electronics manufacturers. High-speed aircraft requiring electronic equipment capable of operating under severe environmental conditions led to the development of a new concept known as CNI (communication, navigation and identification) systems engineering.

Before the CNI systems approach, electronic equipment specifications did not keep pace with new aircraft designs. Consequently, outmoded equipment was often delivered for newly-designed planes. The Collins CNI system was a tailor-made group of standardized electronic modules which provided communication, navigation, and identification functions for a particular type of aircraft. The modules had standardized electrical, mechanical and thermal connections which permitted the same modules to be assembled into different units, as required for each particular aircraft.

By 1966, Collins had installed more than $110 million worth of CNI equipment on more than 4,600 aircraft, including the Republic F-105 and the North American A3J.

Other post-war markets

While military UHF radios and aviation equipment were the primary growth products for Collins Radio in the eight years immediately following the war, other products were also developed.

Sales resumed in amateur radio gear, the original product line of the company. The first Collins post-war amateur radio unit went to its new owner January 8, 1947, during a formal ceremony held in the lobby of the Collins Main Plant. Clyde Hendrix, division president of Pillsbury Mills, Inc., purchased the new 30K transmitter and 75A receiver. The Clinton, Iowa, ham had been one of Arthur Collins' first customers in the 1930s, and placed his order for the new equipment sight unseen.

The 30K transmitter, designed by engineer Warren Bruene, was of entirely new design and incorporated many new features for the ham community. Its companion receiver, the 75A, was one of several units at Collins which resulted from ideas which were being shaped for military applications just before the war ended. The basic idea was to achieve stability and accuracy in the reception of radio signals. The receiver employed an invention called the permeability-tuned oscillator (PTO), which improved frequency control to such close limits that the older method was no longer necessary.

During a demonstration of the new PTO at a ham radio convention in 1947, Ted Hunter, who designed it, was showing the oscillator to a group of engineers. The presentation consisted of independently setting two pieces of equipment to the same dial frequency and then demonstrating frequency separation by means of the resulting beat note. Hunter liked to make the adjustments with the audio output cut

› Lab assistant Kenneth Everhart stands beside a 30K-1 amateur transmitter. Seated is Nobel Hale, advertising editorial assistant, with a 310A amateur band exciter. A color version of this photograph was featured on the cover of the November 1946 issue of *Radio News*.

off, then turned up the output so the beat note was clearly audible, "sort of a blindfold test."

One incredulous individual wanted to try it for himself. He set the independent oscillators, turned on the equipment, heard the beat note and walked away with a vague, puzzled look on his face. Twice more he returned to repeat the test while Hunter stood by trying to hold back a grin. As the puzzled individual walked away from the demonstration for the third time, he was heard to mutter, "By gosh, I guess that ad was right!" Such as the quality of Collins amateur equipment: nonbelievers became believers.

Broadcast equipment

Commercial radio stations enjoyed their golden age during the 1930s, and just prior to the war, development work in television was under way. After World War II, some companies which had previously specialized in radio transmitters turned their attention to television equipment, assuming the pictures and sound would replace the "sound-only" form of communication. Television did enjoy phenomenal growth after the war, but broadcast radio equipment continued to account for a large volume of business for Collins Radio.

Nearly 350 new radio stations went on the air between 1951 and 1955. One of the most numerous types was in the medium power category, with towns as small as 4,000

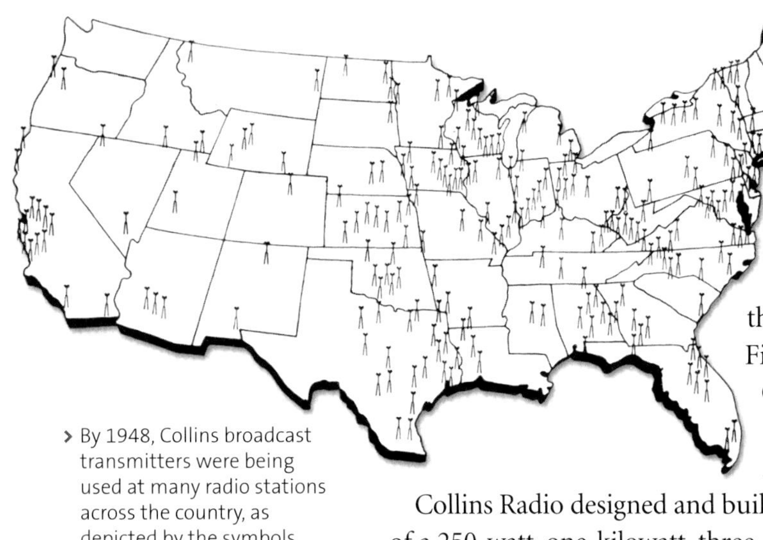

› By 1948, Collins broadcast transmitters were being used at many radio stations across the country, as depicted by the symbols. Hundreds of others employed speech, studio or other equipment manufactured by the company.

population getting their first radio stations. The Collins 20V was one of the best sellers in that category, with sales near 300 by 1955. Lauren Findley was the original project engineer for the 20V. Collins furnished complete AM and FM radio stations, including everything from the announcer's microphone to the station antenna.

Collins Radio designed and built a line of FM transmitters, offering the choice of a 250-watt, one-kilowatt, three-kilowatt, or ten-kilowatt transmitter as standard equipment. At the 1949 National Association of Broadcasters conference, John Green, head of broadcast sales for Collins, presented a paper describing how the Autopositioner could be used for remote control of broadcast transmitters. One year later Collins Radio introduced a one-kilowatt AM transmitter featuring a new circuit design which improved reliability and operation.

Although the company never entered the consumer radio receiver market on any large scale, an effort was made to market a less expensive version of the famous Autotune for home radio receivers. The 496E Autotune was less precise and therefore less expensive than the military and commercial versions, but its price tag proved too much for even the top-of-the-line home radios.

However, another type of broadcast radio receiver manufactured by Collins was successful: a receiver engineered expressly for railroads. The problems peculiar to railroad installations, such as fading signals and interference, were worked out in cooperation with the Rock Island Railroad and the American Phenolic Corporation. The project, which was started in 1945, produced a broadcast receiver and distribution system for clear reception in railroad cars. The system provided reception of any one of ten predetermined broadcast stations, and tape or wire recorders could be used in conjunction with the receiver to furnish a pleasing variety of programs for passengers.

› The control room of radio station WMT in Cedar Rapids contained Collins broadcast equipment.

Another area of broadcast radio equipment for Collins Radio Company in the post-war years was equipment for the State Department's Voice of America network.

In the conflict of ideologies with the Communist block countries, later coined the "Cold War," the weapon was radio. In 1950, the VOA began an expansion program to lengthen its radio time, add 19 languages to its former 26, and increase the effectiveness of its signal. Collins Radio had a big part in equipping the radio ship *Courier* for its debut in the propaganda war. The project, called "Operation Vagabond," was a facet of the "ring plan" designed to ring all of the world's critical areas with extremely high-power communication facilities. The biggest obstacle faced by the VOA was the electrical field created by an estimated 1,250 Soviet jamming transmitters.

Test technicians C. F. Hardenbrook, left, and Hank Rathje consult schematics while "trouble-shooting" one of the large 207B-1 transmitters used by the Voice of America. 1951 photo.

Aboard the *Courier,* Collins field engineers installed two 35-kilowatt shortwave transmitters and a 3-kilowatt transmitter for ship-to-shore communications. An RCA 150-kilowatt transmitter was used for medium wave. Banks of Collins receivers were also used at relay stations to pick up the stateside broadcasts for relay across eastern Europe.

Despite the Kremlin's best efforts, in 1952 the VOA signals could be heard 25 percent of the time inside Moscow, where the jamming was concentrated, and 60 to 80 percent of the time outside the Soviet capital.

Post-war impact

Even though much of the company's employment growth had occurred earlier, Collins Radio had the greatest impact on its host community beginning in approximately 1948.

During the war most of the plant workers either commuted to Cedar Rapids, or were local residents who had previously worked at companies which were not war plants, or were entering the work force for the first time. Because of wartime restrictions, housing construction was at a virtual standstill.

Following the adjustment period at the close of the war, a pent-up housing industry exploded in Cedar Rapids, especially on the north side near the Collins Main Plant. Many commuters moved to town and many new employees moved to Cedar Rapids, presenting a tremendous challenge for home builders, utilities, the school system, and the city.

"There was tremendous expansion in northeast Cedar Rapids and Collins was single-handedly responsible for that growth," said Harold Ewoldt, long-time Chamber of Commerce manager and local historian. "Collins had 'used up' the available work force, then came the boom in housing as workers came in from outside the area. The whole northeast part of Cedar Rapids in the triangle from First Avenue to Collins, plus growth in Marion, is due to Collins Radio."

In 1947, construction began on an 80,000-square-foot building south of the Main Plant to house the finishing department. The paints and solvents used by the finishing department had been causing problems with the air-conditioning system in the Main Plant; the addition also was undertaken to consolidate several scattered departments. ▪

Bottom left: Construction of Building 139 behind the Main Plant as it appeared on October 27, 1947, looking southwest.

Bottom right: The Collins Main Plant area as it appeared from the north in 1950. The tent at the rear of the site was erected to cover the assembly area for a large dish antenna (see Chapter 5). The area at the top of the photograph was used as a trailer park during World War II, but by 1950, most of the trailers had been removed.

New ideas

While much of the post-war research effort at Collins Radio Company was applied to specific products such as UHF military radios and aviation electronics, a separate division was formed to undertake research in general fields of study, without the pressures of contract deadlines and restrictions. Significant scientific developments in electronic communication, atomic research, and flight control came out of this research work at Collins.

‹ Collins engineer Irv H. Gerks tests a milling machine by shaping a small parabolic "dish" antenna. The machine was later used to shape a 50-foot-diameter dish antenna.

› In 1947 the government decided it would be necessary to assign UHF channels for television broadcasting due to a shortage of VHF channels. At this time they were considering adapting Resnatron for television transmitter purposes.

Starting in 1946, Collins had under study, both independently and in conjunction with government agencies and labs, new methods of reliable, long distance communication. Out of this investigation emerged what was known as Transhorizon communication techniques: methods by which VHF and UHF signals were transmitted beyond the line of sight. Many of the achievements in Transhorizon techniques were outgrowths of advanced work with a high-power transmitting tube known as the Resnatron.

The Resnatron principle was originally developed in 1938 by Winfield Salisbury (Cedar Rapids native and friend of Arthur Collins) and two associates at the University of California. The scientists were attempting to find a solution to the problem of obtaining high power outputs at extremely high frequencies. Like so many other scientific projects, the Resnatron was pressed into military service during World War II. An urgent need arose, in connection with the radar countermeasures program at Harvard's radio research laboratory, for vacuum tubes capable of generating high power levels at frequencies in the range from 350 to 650 megacycles. This program culminated in the development of the Resnatron, which was used to jam airborne interceptor radar equipment carried by German night fighters over the English Channel.

When it became evident in 1947 that it would be necessary to assign UHF channels for television broadcasting due to a shortage of VHF channels, the government considered the possibility of adapting the Resnatron for television transmitter purposes. In view of these circumstances, Salisbury was hired as director of research at Collins to continue his Resnatron studies. Much of the initial experimentation was performed by S. G. McNees, W. J. Armstrong, and Roger Borne of the research division staff.

> Dr. Walter Kohl (right), head of the vacuum tube laboratory at Collins Radio, and Dr. John Clark, senior engineer, discuss the new Resnatron tube which the company constructed.

To demonstrate the potential of the Resnatron for a variety of uses, Salisbury had the equipment rigged to produce microwaves of such intensity that they could cook a hamburger in 30 seconds or pop a sack of popcorn almost instantly. He demonstrated his "electronic oven" at the 1949 Iowa State Fair as one of the centennial exhibits in the Des Moines Register and Tribune building, and the device received much attention. Collins Radio, however, never developed a production model. Other companies, notably Raytheon, developed microwave ovens of their own design.

For Collins Radio, the more serious applications of Resnatron research dealt with electronic communications.

Faced with the difficult problem of deciding how far apart to place UHF stations on the same wavelength, the Federal Communications Commission selected Collins Radio to make the measurements because of the availability of its Resnatron equipment at its laboratory at the Cedar Rapids airport. DuMont Laboratories, CBS and RCA also investigated other aspects of UHF broadcasting.

At Collins, Irvin H. Gerks, nationally-known radio propagation expert, headed field studies at a number of remote locations approximately 100 miles from the Resnatron transmitter at the airport. Something unanticipated, and "rather startling," according to Gerks, was the discovery that on some summer mornings the signal received from Cedar Rapids seemed to be well within line-of-sight range, even though the portable receiving station was positioned well beyond the curvature of the earth.

But as the day wore on, the signal arrived weaker, presumably from scattering in the lower portion of the atmosphere, called the troposphere. The discovery led to development of a new form of radio communication known as scatter propagation.

Another area of radio research dealt with cosmic rays, which were microwaves emanating from the sun and other celestial objects. In 1949 Collins researchers built a heavy power field station at a site called the Feather Ridge Observatory, situated along the Palo road near the Crawford stone quarry northwest of Cedar Rapids. One of the

first experiments conducted at Feather Ridge was observation of a total eclipse of the moon through radio telescopes — the first time this had ever been done. From this experiment, Collins engineers observed that the lunar microwave temperature did not differ greatly from periods of direct sunlight on the moon's surface. This led to the conclusion that lunar radiation came not only from the surfaces, but also from layers of materials beneath the surface — far enough below the surface that the temperature remained constant.

The Feather Ridge studies, by Dr. Dale McCoy and C. M. Hepperle, led to construction by Collins of a giant aluminum dish antenna for the Naval Laboratory at Anacostia, D.C., in 1950. The 50-foot parabolic receiver, which was then the largest ever built, was used by the laboratory in its investigation of radio waves originating from celestial bodies.

"It has long been known that there is a connection between the activity of the sun and long-distance radio communication between any two points on the earth. To further this investigation, it has become necessary to use increasingly larger antennas in order to achieve more sharply directive beams which enable one to locate with precision the sources of radio waves arriving from outside our atmosphere," Salisbury explained in a news release.

The deep space radio telescope dish was made in 30 sections, each weighing about half a ton. They were cast for Collins by the Aluminum Company of America. The sections were bolted on a 35-foot circular girder and placed with steel dowel pins to permit disassembly for shipment. The assembly and machining of the dish were done in a tent south of the Main Plant because there wasn't factory space available for such a large project. Don Holzschuh, engineering services department, said the milling head cut off about two tons of metal chips as the reflective surface was machined. Model fabrication mechanics assigned to the project were C. E. Carlson, G. W. Eiben, L. Meyer, H. H. Harris, P. G. Scott, and H. J. Tiedemann. The axis converter, which directed

› The vacuum tube laboratory, part of the Collins research division, worked with new designs for high-power transmitting tubes. The man at the glass lathe is John Barton.

› Dr. Gene Marner with a radio telescope at the Feather Ridge facility.

Wingless aircraft and flying boats

> This full-scale "test bed" of the Aerodyne never flew, but was built to test the design principles of Dr. Lippisch's unique flying machine.

One distinguished scientist to come to Collins Radio Company in the post-war period was Dr. Alexander Lippisch.

His Delta I glider, built in 1930, was converted into a powered plane and shown in flight to the public in 1931. Lippisch further developed his idea and designed the first high-speed rocket-powered aircraft, the ME 163 Komet, which flew 625 miles per hour for the German Luftwaffe in 1941.

A model of his swept-back Delta glider, together with results of his research, showed the superiority of this type of high-speed aircraft, used today in many jet fighter designs.

As Nazi Germany collapsed in the spring of 1945, the United States raced in to grab as many of the highly skilled German scientists as possible. Under code name "Operation Paperclip," Dr. Lippisch was one of 50 German scientists brought to the United States.

Dr. Lippisch joined Collins Radio in February, 1950, as head of aerodynamical research. It was here that he developed the Aerodyne, an unusual wingless aircraft. The Aerodyne project, funded by the Office of Naval Research, took place in the Collins Aeronautical Research Laboratory at the Cedar Rapids airport, and ultimately a full size non-flying "test bed" was constructed. The test model wasn't intended to fly, but the Navy insisted for continued funding that it must fly, so an engine and propeller were added. The craft was shipped to Moffet Field near Sunnyvale, California, for wind tunnel testing, and there the project was scuttled in 1962.

At the Collins Aeronautical Research Lab, Dr. Lippisch also developed an advanced smoke tunnel to study the flow of air over airfoils.

At the conclusion of the aerodynamic studies at Collins, Dr. Lippisch's efforts switched to another project initiated by Arthur Collins — boat hull research.

Collins was looking for ways to improve efficiency in planing hulls.

In 1959, aerodynamics lab personnel were moved to a new laboratory in the Butler Buildings near the Main Plant. The test area was equipped with a 90-foot-long, 12,000 gallon tow tank for testing scale models of new hull designs. A glass window in the bottom of the tank permitted photographs to be made of the flow pattern which resulted

> An interceptor drone prototype designed by Dr. Alexander Lippisch.

> The Aeroboat was flown successfully several times at the Coralville Reservoir by pilot Clayton Lander.

from movement of the model through the water. A two dimensional flow tunnel — one of the first of its kind — was constructed to investigate flow without the complications of side or end effects.

Five to six persons performed research at the marine lab, while a group of 15 persons constructed models and full-scale boats.

From 1960 to 1962, several boats were built at the lab, and operational tests were performed on Cedar Lake and the Cedar River. The design efforts in Cedar Rapids culminated in 1962 with construction in Newport Beach, California, of a 72-foot fiberglass boat, now owned by Arthur Collins.

The advanced designs produced at the marine lab were never adopted by the boat manufacturing industry, largely because the deep "V" hull was designed at about the same time, and provided good turning and handling characteristics along with good performance at high speeds.

In 1963, the Collins marine lab, under the supervision of Dr. Lippisch, designed and built the Aeroboat. The craft was the first ever designed to use the ground effect principle over a body of water. The wooden experimental craft was originally licensed as a boat, but after a few tests, it became apparent that it would fly as an airplane. Test pilot Clayton Lander called the FAA and told them the Collins Aeroboat might need to be reclassified as an airplane.

> Dr. Lippisch, seated at control console, demonstrates a flying model of the Aerodyne to a tour group at the Collins hangar. The model used electric motors to force air downward to create lift.

"When do we become an airplane?" Lander asked the FAA officials. They studied the question for several days and reported back to the Collins group that if it flew more than 28 inches off the surface of the water, it should be reclassified as an airplane. Since the Aeroboat had flown more than 100 feet high, it became an airplane.

The original wooden craft is now part of *Flight and Fancy — The Alexander Lippisch Collection* at the Carl and Mary Koehler History Center in Cedar Rapids. After Dr. Lippisch's retirement from Collins in 1964, he worked with other firms to refine Aeroboat principles. Dr. Lippisch died in 1976 at age 81. ▪

> The two-dimensional smoke tunnel designed by Dr. Lippisch was acclaimed as one of the finest available for studying airflow over airfoils.

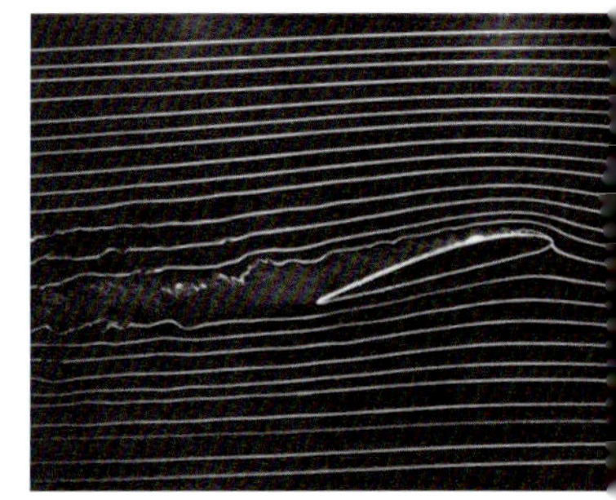

› The 50-foot diameter parabolic antenna was assembled atop a naval laboratory building in Anacostia, D.C., and was used by the Navy to study deep space radio emissions.

movement of the dish to follow celestial bodies across the sky, was the design of Irv Gerks, John Ramson, and Ross Pyle.

In June, 1951, Collins Radio constructed a special propagation facility consisting of a building to house several high-power transmitters. The site, located near the Cedar Rapids Airport, was known as the Konigsmark Laboratory. Experiments at the lab were in frequencies in the VHF band. All operations from this site was devoted to studying "ionospheric scatter," a form of radio transmission which used the high-altitude ionosphere to scatter a strong radio signal so that a portion of the signal made its way back to the intended receiver.

Since the signal power received over long VHF Transhorizon circuits was found to be very low at times, it wasn't suitable for efficient teletype transmission. Collins researchers had been experimenting since 1949 with a signaling system which represented a radically different approach to the teletype transmission problem. They conceived the Predicted Wave System which provided greatly improved long-range teletype transmission reliability. It made it possible to put 40 channels in one standard three-kilocycle bandwidth, each capable of transmitting 100 words per minute.

National attention was drawn to another research project in November, 1951. Scientists from Collins and the National Bureau of Standards used the moon as a giant reflector to bounce UHF signals from Cedar Rapids, Iowa, to Sterling, Virginia.

"The experiments indicate the feasibility of using the moon as a natural relay station for radio communications between two points on the earth's surface," explained Arthur Collins. Specially constructed for the experiment was a 75-foot "tapered wave guide horn" antenna, installed at the Collins hangar at the Cedar Rapids Airport. Shaped like a huge funnel, the antenna directed the radio signals to the moon. On the trip back to earth the signals were received by a 30-foot dish reflector installed at Sterling by the National Bureau of Standards. Using a 20-kilowatt transmitter with power generated by the Resnatron tube, the coded signal was sent the one-half million miles from earth to moon to earth. When the signal was decoded in Sterling it revealed a message that had been heard in the vicinity before. It was the historic message, "What hath God wrought!," used by Samuel Morse in 1844 over his new telegraph line from Washington to Baltimore.

› To keep the heat of the sun's rays from warping the precision-machined aluminum sections, a 34-foot high tent was placed over the dish antenna assembly area.

› At the Konigsmark Laboratory, four rhombic antennas ranging in size from 500 to 2,000 feet in overall length were constructed in 1951. They were connected to several high-power transmitters housed in the metal building near the center of the photograph. The Cedar Rapids Airport can be seen in the background.

› Collins Radio personnel explained the functions of the Konigsmark antennas to a group of U.S. Air Force representatives in August, 1952. From left: John Shanklin, J. Chisholm, Arthur Collins (Collins Radio); H. Lindner, R. Kaplan (U.S.A.F.); R. McCreary (Collins Research Division director); and H. Wells (U.S.A.F.).

The Collins Cyclotron

In 1947, under a contract with the Atomic Energy Commission, the first commercially-built cyclotron began as a project of Collins Radio Company. For the first time, construction of an "atom smasher" was put on a production budget and certain standards of performance were guaranteed. The accelerator was delivered to the AEC's Brookhaven National Laboratory on Long Island.

When it received the $750,000 cyclotron contract from the AEC, Collins Radio was the first firm to enter the field. General Electric was the only competitor. There were many such devices in use in the United States and other countries, but all were hand-made and no standards of performance had to be guaranteed. Manufacturing a cyclotron on a production schedule was something new. The job had to be done with the skills and tools of a private concern rather than the manpower and facilities of the large universities.

The high energy machines were regarded as holding great prospects in the field of medicine for cancer therapy and in the field of energy for development of atomic power sources.

The most bulky item in the Collins cyclotron was the 250-ton magnet, cast by the Carnegie-Illinois Steel Company. The coils were wound by Collins workers at the main plant. Aluminum stripping as wide as a man's palm was used. Problems arose in assembling the "dees," the copper vacuum chambers in which the atomic fragments spin. When the massive tubes were placed in the Collins 250-ton hydraulic press to flatten out the ends to form the vacuum chamber, the jaws of the giant press just bounced. Collins technicians finally rigged up an arrangement of ganged levers, powered by a hydraulic jack, which squeezed the ends to the desired shape. Welding copper also proved to be a difficult chore. Collins developed a technique using helium and a tungsten arc which did the job.

Atomic particles injected into the machine's center were accelerated by electric forces and steered by the magnet into a spiral path which brought them to the target area. The specimen target, being bombarded by particles, was made radioactive. In other words, the cyclotron gave high speed to particles and used them as projectiles for nuclear disintegration.

One of the outstanding features of the Collins cyclotrons was the large number of high-speed particles they produced — about a million billion per second. Expressed in terms of conventional current units (for each particle carried a charge of electricity) the output was 200 millionths of an ampere.

A second cyclotron, very similar to the one made for Brookhaven, was completed in 1952 by Collins for the Argonne National Laboratory at Lemont, Illinois near Chicago. The facility was designed, constructed, installed, and adjusted to full performance by Collins Radio at a cost to the lab of $966,000. More than a dozen scientists and technicians, under the direction of Dr. J. J. Livengood, accomplished the second and final installation of a Collins cyclotron.

In October, 1951, Dr. Livengood announced his resignation from Collins to accept a position at the Argonne Laboratories, to direct use of the cyclotron. ▪

› Winfield Salisbury displays a model of the atom-smasher built for the Atomic Energy Commission at Brookhaven National Laboratory on Long Island.

> Top: Two exciter coils near completion for the 250-ton magnet of the Brookhaven cyclotron. Each coil contained nearly three miles of hollow, water-cooled aluminum strap.

> Left: The control room of the Collins cyclotron at Brookhaven.

> Right: The completed cyclotron at the Brookhaven National Laboratory.

› An early version of the Collins radio sextant, at the Feather Ridge Observatory.

The Collins-NBS experiment was the first time a long distance message had been sent via the moon, and was the first use of UHF frequencies for this purpose. (Six years earlier the Army Signal Corps successfully bounced an unintelligible radio signal off the moon.)

In 1952, Collins installed an experimental communication link between its Cedar Rapids Airport laboratory and its new Dallas laboratory. (*See Chapter 6*) The link served as a field test facility for the Collins Predicted Wave Teletype system in long-range point-to-point service. In addition, the Cedar Rapids-Dallas link was used for research and investigation of various modulation techniques including the comparison of voice transmission by narrowband FM and a system called single sideband.

Two years later, Collins established a tropospheric scatter circuit between the engineering building in Cedar Rapids and a terminal near Lamar, Missouri. The Lamar terminal was located on a direct line halfway between the Cedar Rapids terminal and the Dallas facility.

Collins Radio gained the attention of the news media in April, 1950, when the *New York Times* reported on the Transhorizon experiments being conducted by Collins. The feature-length article described the discovery of "a new way of sending radio signals through the air that holds the promise of revolutionizing long-distance communication, and conceivably might open the door to international television. The new method appears to make obsolete the generally accepted theory that signals transmitted on very high frequencies, such as those used by video, are limited to line of sight."

The Collins researchers discounted the possibility of using their techniques for international television because of signal quality problems. But they agreed that Transhorizon techniques could revolutionize long-distance communication.

More experiments and achievements followed. In 1953 Collins designed and constructed a 50-kilowatt biconical and pyramidal antenna for the Air Research and Development Command at Prospect Hill, New Jersey. The 8,000-pound aluminum antenna was used in research work of the Air Force's Cambridge Research Center.

This pioneering work enabled Collins to have a complete line of Transhorizon equipment ready in the mid-1950s when the Distant Early Warning (DEW) Line was established.

With Russia's first explosion of an atomic bomb in 1949 and the invasion of South Korea in 1950, American leaders realized the need for an electronic early warning system to detect intruding aircraft approaching through the Arctic. To be effective, the DEW Line had to coordinate its many parts. Conventional high frequency communication equipment dependent upon ionospheric refraction could not be relied upon because of the frequent blackouts and disturbances in the Arctic caused by magnetic storms, aurora borealis, sun spots and other phenomena. Land lines or microwave relays were out of the question in this Arctic desert because of the cost.

The answer came with the development of Collins Transhorizon, or scatter, communications. By means of high-powered transmitters, high-gain antennas and sensitive receivers, Transhorizon systems were able to use the scatter effect in VHF and UHF radio wave propagation to achieve highly persistent communication beyond

› Collins Radio project engineer Ted Willis discusses features of the radio sextant with Capt. John Brandt and Lt. John Kuncas of the U.S. Navy. The instrument was installed aboard the *USS Compass Island* in 1959.

U.S. Navy photograph.

the horizon. The equipment was used to tie the DEW Line with the continental defense complex, and to connect the DEW Line stations with each other.

Behind the scenes of these publicized developments, a similar project was kept under security wraps. Dr. Dale McCoy, who was instrumental in much of the cosmic ray antenna research, designed a new navigation device based on those principles.

Within a year after publications appeared in 1944 and 1945 describing weak microwave radiation emitted by the sun, Collins began research into radio astronomy, and shortly afterward investigated the feasibility of an all-weather sextant, an electronic counterpart to the hand-held optical sextant which depended on a clear sky. Most of the research was conducted at the Feather Ridge facility. Dr. McCoy, R. M. Ringeon, and C. M. Hepperle were largely responsible for the development of the world's first radio sextant.

While the optical sextant used direct light waves from the sun to determine the ship's position, the Collins radio sextant used radio waves from the sun. The sun's radio waves easily penetrate the clouds, enabling all-weather operation of the radio sextant in the daytime.

Announcement of the radio sextant in the summer of 1954 again drew nationwide attention to Collins Radio. Captain P. V. H. Weems, chairman of the board, Aeronautical Services, Inc., visited Cedar Rapids shortly after the announcement. He commented, "The opportunity to inspect and test the original Collins automatic radio sextant has convinced me that we are entering a new and extended phase of practical celestial navigation."

Time magazine, in an interview with Fred Haddock, radio astronomer of the Naval Research Laboratory, reported: "The ship's navigator can find his position just as if he had an assistant watching the sun through an ordinary optical sextant. No cloudy weather gets in the way of the radio sextant, nor can an enemy jam the radio impulses (as is possible with other radio aids to navigation, such as Loran)."

The first radio sextant detected only energy from the sun, which restricted its effectiveness to daylight hours. However, as advances were made and a more sensitive antenna and receiver were developed, a sextant was built which could track both sun and moon. ▪

Growth in the 1950s

The decade of the 1950s was a period of spectacular growth for Collins Radio Company, and the electronics industry in general. In 1950, total company employment was about 2,000. By 1959 that figure had reached nearly 11,000. Annual sales of less than $13 million in 1950 soared to $118 million by the end of the decade.

Collins received growing recognition in all its markets, especially in the burgeoning areas of microwave communications, single sideband, and aviation electronics.

‹ Much of the steel framework was in place by June 1953 for the new engineering building (Building 120) in Cedar Rapids. The unpaved road at the bottom of the photograph is C Avenue.

Writing in the New York edition of *Commercial and Financial Chronicle*, Philip Carret, a New York Stock Exchange broker, assessed the condition of Collins Radio Company at the beginning of the 1950s:

"As a consequence of this high regard for the company's products, Collins' backlog, which has risen $100 million in the past year, continues to grow. It is currently reported to be in excess of $150 million. During World War II the rate of shipments was $5 million a month. At the presently increased level of prices and with greatly expanded factory facilities, a higher level of output can obviously be expected.

"The Collins management does not consider that this huge increase in business on the books is merely a reflection of the rearmament boom. A far-sighted policy of emphasizing research and development was adopted after V-J Day and a considerable part of the current backlog is attributable to demand for improved and wholly new products which would have been ordered even under normal conditions. The size of the present backlog is a tribute to Collins' engineering, research, and far-sighted management. However, there is no reason to believe that the backlog has by any means reached its peak. Air Force requirements are steadily being expanded and the trend in electronic equipment is toward more complex and expensive equipment."

Starting in 1950, Collins Radio Company undertook a five-year, $5 million program of fixed asset investment, which included new laboratory and manufacturing facilities. In the first two years of the program, the company doubled its floor space, and more expansion followed.

> The Cherry Building was located at 317 10th Avenue SE in Cedar Rapids.

Part of Collins Radio's new research division was moved in the summer of 1950 to a leased three-story building, formerly the site of the Cherry-Burrell Corporation's factory in Cedar Rapids. The large structure was also used as a warehouse for overflow and obsolete stock.

In April, 1951, the purchasing department moved from the Main Plant to a building on Second Avenue near Fifth Street in Cedar Rapids. The purchasing department occupied the lower two floors, and the upper floor housed a training school for military radio men. The industrial engineering department took over the area at the Main Plant left by the purchasing department.

National business publications reported that the electronics industry was due for vast expansion. The military procurement logjam had broken wide open, and the government agreed to bigger tax deductions for some manufacturers who were turning out products for the armed forces. Those firms included big names such as RCA, Westinghouse and General Electric, and specialty manufacturers such as Federal Telephone & Radio and Collins Radio Company. Uncle Sam began to string radar stations around the United States and Canadian coastlines, airfields were reactivated, and naval vessels were pulled out of "mothballs." Modernized tanks and other vehicles of modern warfare used radio and other electronic equipment. The Pentagon wanted the electronics industry able to multiply its output in case another big war started.

With increasing military orders, Collins was urged by the Defense Department to consider further decentralization of its facilities for security reasons. Management began studies of a number of communities. Collins Radio announced in May, 1951, an expansion program to build a $1 million plant near the Dallas, Texas, suburb of Richardson. Another Dallas building was leased to start mechanical assembly production while the new plant was under construction. Collins also announced plans to lease a hangar at nearby Redbird Airport to install and repair airborne equipment.

> For a time, women "white room" employees at the Main Plant wore head coverings called "snoods" to keep hairs out of delicate flight instruments. But some managers thought visitors got the erroneous impression that only nuns worked in the white room, so the snoods were replaced with caps. 1952 photo.

Most all employees for the plant came from the Dallas area. James Flynn, Jr., was named general manager. Flynn came from American Airlines where, as superintendent of communications, he ordered Collins Autotune equipment for the American Airlines fleet in 1937. By June, 1951, three other managers had been assigned to the Dallas facility. W. G. Pappenfus was named director of manufacturing, Harold Moss was made manager of test and inspection, and Arthur Luebs was named senior buyer for the Texas Division.

"Our decision to locate the new plant in Texas is in line with the current practice of separating production plants geographically for security reasons," Arthur Collins explained. "So long as we are locating another plant away from our main operation, we picked a place close to the heart of the aviation industry, and where the weather would give us more uniform test flight conditions. We found exactly the conditions we were looking for in Texas."

The new one-story, 50,000-square-foot plant was similar in design to the main plant in Cedar Rapids but was about half the size.

With the announcement of the Dallas facility, the company also made plans to expand operations at the regional sales office in Burbank with another 10,000 square feet and additional employees, and to add new equipment at the Main Plant and Cherry-Burrell building in Cedar Rapids. The federal government granted Collins a five-year tax writeoff for about $800,000 of the projects.

Actually, the company at first did not intend to expand the Burbank plant. Collins had originally announced in December, 1950, plans to build a new $500,000 plant in Arcadia, California for research and manufacturing. The Arcadia Chamber of Commerce and a local citizens' committee wanted the Collins plant in their town, but a vocal group of Arcadia residents opposed the new plant. They presented a petition which said a new Collins plant would set a precedent for more factories in the predominantly residential suburb of Los Angeles, would provide a strategic bombing site, and would increase traffic and smog. The Arcadia City Council voted three to two to turn down the Collins request for rezoning, so the plan was dropped in January, 1951. Instead, the Burbank facility was expanded and made headquarters of the new Western Division of Collins Radio Company.

The first contract for the Western Division, located at 2700 West Olive Avenue in Burbank, was a testing job for the Atomic Energy Commission. The first production job and the beginning of an assembly line was for the construction of servo amplifiers and drive units for Lockheed Aircraft Service. Later projects included guided missile receivers, a navigation trainer for the Navy, a voice/code reproducer for the Civil Aeronautics Administration, and general production overflow from the Main Plant in Cedar Rapids. A major function of the Western Division was to aid aircraft manufacturers in the California area in determining their radio and electronic needs, and to service Collins equipment in that area. Carl Service, who had previously been Collins Radio's service manager on the west coast, was made the first Western Division manager.

> Doris Gilbert Werth (front) and Edith Atkinson set type on the Collins publications department's Justowriter machines. The department was located in the Third Street Building. 1957 photo.

> Part of the Texas Division was located in this leased building on Hi Line Drive in Dallas.

New engineering building

In May, 1952, the company submitted an application to the federal government for a certificate of necessity covering construction and amortization (for tax purposes) of a new engineering building in Cedar Rapids (now called Building 120). The certificate was approved that autumn, and Collins announced that a $1.8 million engineering laboratory was to be built to house approximately 600 employees in 12 labs and additional office space. Other space was to be provided for a lobby, technical library and a cafeteria-auditorium. The site was a 52-acre wooded tract at the intersection of Old Marion Road and C Avenue, just northeast of Cedar Rapids. At that time, engineering activities in Cedar Rapids were scattered at a number of locations. The new plant was designed to concentrate most engineering at a single site.

An existing mortgage-loan agreement was extended to finance construction. Under terms of the agreement, Collins Radio put up as collateral real estate and buildings owned by the company in Cedar Rapids as well as machinery and equipment. The Linn County Board of Supervisors paved the way for construction in October, 1951, by approving a zoning change from residential to light industrial. With all formalities cleared, Collins exercised options it held on the land, and Marie Carver and her son, Weston, sold the tract to Collins.

The following March, the City of Cedar Rapids expanded by 52 acres when it annexed the site where the new building was under construction. Speaking for Collins Radio, attorney C. J. Lynch noted that city officials had proposed the annexation, and Collins felt the decision was fair, since the company realized the area would become an integral part of corporate Cedar Rapids. C. W. Garberson, city attorney, pointed out that the city would benefit with many citizens employed at the laboratory, and that its value would provide added tax revenue to the city.

"The attitude of Collins officials with respect to paying for services and taxes has been so fair, and the general type of industry is so outstanding, that the city should take every step it can to extend the services they need," Garberson said.

> The first Western Division production facility was located in Burbank, California.

› The site of the new engineering building was directly east (left) of the grove near the intersection of Old Marion Road (horizontal in the middle of the photograph) and C Avenue (upper left to lower right). Collins Road was not yet constructed in 1952. The Collins Main Plant can be seen in the upper right corner.

› Arthur Collins turned the first chunk of frozen ground in a ceremony to start construction of the engineering building in January 1953. Pictured from left: C. G. Selzer, J. B. Tuthill, R. S. Gates, Bob Weinhardt, W. H. Bigger, E. D. Broderick, and Collins.

One feature in the construction was the addition of wire screens in all the concrete walls to keep radio signals from leaving the building. Some residents in the Main Plant neighborhood had complained of interference with their television sets, so Collins took measures to eliminate the problem. Most of the problems at the Main Plant arose when large Voice of America transmitters were tested.

The big move to the new engineering building started in November, 1953, and continued through March, 1954. All major segments of the move were done on weekends, with planners achieving their goal of losing no more than one day's work in each department moved. Working with plant engineer Bill Weinhardt were Walt Brown, who was in charge of moving offices, stock rooms and the library; and Merrill

> The engineering building under construction in 1953.

Ludvigson, who supervised the moving of lab facilities. Each weekend vans were at the building at 5:30 p.m. Friday to transport material to the new location. The first day would end at 1:30 a.m. Saturday with the material in its new location and placed in its new space. At 7:30 a.m. Saturday the crews returned to unpack, clean and arrange the material, an operation usually completed by 1:30 p.m. The electrical wiring and telephone service connections were also done on the weekends.

All major departments in the engineering division except one were moved. The moving company, Calder's Van and Storage Company of Cedar Rapids, was the same one which, in 1933, moved Collins Radio into its first factory at 2920 First Avenue using two small truck loads. Twenty years later, more than 125 vans full of equipment and furnishings changed location — to move just the engineering division.

Construction and moving weren't the only large tasks resulting from the new engineering building. Landscaping received much attention, since the building was to be situated in a natural grove of trees.

> The engineering building was complete and occupied by mid-1954.

First to get the attention of Earl Fredrickson and his grounds maintenance crew were the 181 trees already on site. Thirty-two were stragglers which were removed, along with 13 big trees beyond repair. The other trees required major surgery by the tree service working with Collins maintenance people. Special feeding of the trees worked their roots to the proper level for grading, and every tree was cabled together to prevent limbs from splitting off in windstorms. Grading was difficult because of the many trees, so much of the work was done by hand. In the fall, the crew planted 3,000 pounds of grass seed. Evergreens from all parts of Iowa and Illinois were planted around the entrance of the building. Fredrickson said this was the first full-grown planting of such size in this part of the country. The evergreens weighed up to seven tons and their transportation was a problem in itself.

"No effort was made to rush through the job," Fredrickson said. "We are anxious to do the best possible work, with the results a 52-acre park of which the employees and the city will be proud."

Two subsidiaries formed

Collins Radio Company of Canada was founded in 1953 as a wholly-owned subsidiary of the parent company in Cedar Rapids. An office was established in Ottawa to provide technical assistance to a Canadian manufacturer of Collins-designed equipment, to maintain liaison with the Canadian government and to promote sales of Collins products.

In 1954, the headquarters moved to Toronto and operated from a hotel room for several months to lay the groundwork for production facilities. Collins of Canada leased a building in February, 1955, and production got under way at the new plant the following July. Key product lines in the early history of Collins of Canada were UHF radios for the Canadian government and the DEW Line. With expansion of the original building and the leasing of a second building, total floor space reached 50,000 square feet by 1956. The Toronto plant also housed a research and development lab, Collins-Canada sales headquarters and a field service repair depot.

> The Collins of Canada building on Bermondsey Road in Toronto.

A second subsidiary was formed in January, 1956, when Collins purchased Communications Accessories Company of Hickman Mills, Missouri (17 miles from Kansas City). The Collins board of directors issued 33,150 shares of Class B (non-voting) common stock for all the stock of the Missouri company and acquired full ownership.

Communication Accessories Co. was founded in 1948, and sales increased steadily from $15,000 its first year to nearly $1 million by 1954. The firm employed 450 persons in three shifts using 20,000 square feet of floor space to manufacture advanced toroidal coils, magnetic amplifiers and filters. Government contractors, who used many of the company's components in missile work, were the largest single customer category, purchasing about 60 percent of the firm's output.

Typical CAC customers included Collins, Western Electric, Western Union, Motorola, Bendix, RCA, Westinghouse, General Electric, Crosley, and Wilcox Electric Company.

Like Collins Radio, the Missouri company was successful because of a significant invention which placed it apart from its competition. The invention was a device that could wind coils on cores extremely rapidly. And also like Collins, CAC was founded by a man who carried through an idea and built a working model.

Ed King constructed his first coil winder in the basement of his father-in-law's grocery store in 1948. King also originated a process to encapsulate toroids in plastic. After the purchase of CAC by Collins in 1956, King remained as president and general manager of the subsidiary until 1959, when he left to form King Radio, which now competes with Collins in the aviation electronics marketplace.

In 1957, CAC moved into its new 57,000-square-foot plant at Lee's Summit, Missouri. The 1957 Collins Radio annual report cited the move and a ten-day strike at CAC as major factors which caused a $235,000 loss for the subsidiary that year.

One month after CAC moved to Lee's Summit, a tornado struck Hickman Mills, the former area of operations for the subsidiary and where many CAC employees still lived. No CAC employees were seriously injured, although homes of some were destroyed.

In 1962, the operations of CAC were moved from Lee's Summit to California and combined with the components manufacturing effort at Santa Anna to form

Amateur equipment — The sentimental favorite

While Collins Radio's product line diversified, its oldest product line — and the sentimental favorite of veteran Collins employees — was the famous Collins "ham" line.

One of the biggest flurries of interest on the amateur radio frequencies since the introduction of amateur single sideband equipment in 1955 was created two years later by the announcement of the KWM-1. This small mobile ham transceiver provided the first opportunity for the mobile operator to benefit from single sideband. Collins engineers studied interior sketches and measurements of a variety of automobiles, then followed up in new car showrooms and in the company's parking lots.

Within the first week of selling, more than 500 orders were taken by telephone, sight unseen, delivery dates and prices undetermined.

When screaming mobs in Venezuela besieged Vice President Richard Nixon and his party in 1958, immediate communication was urgently needed with Washington, D.C. Phone lines leading out of Caracas were completely tied up by the crisis, so Nixon's pilot and longtime ham, Colonel Tommy Collins (no relation to the company) rushed to his hotel room, grabbed a battered suitcase, flipped open the lid and removed a Collins KWM-1 mobile transceiver. Dropping an antenna out his hotel window, the colonel fired up the set and in minutes was in direct contact with the White House through phone patches by American amateur radio operators.

The successor unit to the KWM-1 traveled with many explorers to remote regions of the earth. The best-known explorer to use a KWM-2 was Sir Edmund Hillary, the New Zealander who first conquered the world's highest peak — Mt. Everest. It was during Sir Hillary's 1960 expedition that the Collins KWM-2 transceiver won its stripes as a mountain climber. And more importantly, the equipment played a vital role as a lifesaver on the expedition.

Tragedy first struck the expedition above the 20,000-foot level when Peter Mulgrew, the radio operator, began hemorrhaging from the lungs. His companions slipped and clawed their way back down the mountain with Mulgrew in a make-shift stretcher. It took them three days to reach the 19,000-foot level where the radio was.

Firing up the KWM-2, the explorers raised Katmandu and asked for a helicopter rescue. It took another three days to lower Mulgrew to a point where a helicopter could reach him and fly to a hospital.

Shortly thereafter, Sir Hillary was stricken with a mild heart attack and the expedition again signaled for assistance.

Although both of Mulgrew's legs were amputated because of frostbite, he continued to operate a ham rig during his convalescence — the same set that traveled with him and transmitted the lifesaving message.

In 1959, Collins introduced the S-Line, a new line of amateur gear. Advanced circuit design simplified operation, and the line featured compact, functional styling.

The next year, one of the most devastating earthquakes on record ripped through Agadir, Morocco, killing 10,000 inhabitants and injuring more than 35,000 others. First to bring relief was a team of Military Affiliated Radio System (MARS) operators equipped with Collins S-Line. During the four days following the disaster the MARS team provided most of the communication in and out of the stricken city.

Why is Collins ham radio gear the noted leader? There is no single reason. The Collins amateur product line traditionally had the advantage of top engineering talent. Another factor is the company's concern with mechanical and physical design characteristics. The manufacturing and testing procedures which Collins amateur gear goes through is another reason for the equipment's "second to none" rating. A final factor is the sales and distributor forces developed for Collins amateur products.

"Combine all these components," as one sportsminded ham said in 1961 when asked to evaluate Collins amateur equipment, "it's like the New York Yankees, War Admiral and Joe Louis of ham equipment all packed into one box." ▪

› An original advertisement for the Collins KWM-1 touts the radio as "the finest mobile rig available."

> Top: The mobile KWM-1 amateur radio is demonstrated by John Hunt in a 1958 Chevrolet.

> Bottom left: S-Line "ham" units were the neat and attractive radios considered to be top-of-the-line equipment for amateur operators.

> Bottom right: In 1979, the Collins KWM-380 became the first amateur radio introduced by the company since the S-Line units 20 years previously. Dave Berner, amateur products program manager, watched as Al Dorhoffer, editor of *CQ* magazine, operated the new transceiver.

the company's Components Division. Certain administrative functions of the new division were combined with those of the Information Science Center at Newport Beach, California. (*See Chapter 8.*)

1956 fire

A defective oil heater set off a blaze causing a $200,000 loss in two metal buildings at the Cedar Rapids Main Plant February 6, 1956. Originating in one of the Butler buildings used for maintenance and as a garage, the fire destroyed maintenance stock, tools and a new station wagon. In the adjoining components test lab, heat from the fire damaged test equipment and offices.

Within minutes after Howard Batchelder and Floyd Ladman were forced from the burning building, brigade members and other night shift volunteers rushed coatless into the subfreezing weather to fight the fire. A two-alarm call brought city firemen from five companies. Flames were subdued within a half-hour, but heavy smoke poured out of the buildings for some time.

Prompt action by plant guards, Collins fire brigade personnel, maintenance men and volunteers won the praise of company officials. ▪

Microwave communications

A significant field which unfolded in the early 1950s was the development of microwave communications.

Microwaves are ordinary radio waves, but are extremely small in wavelength. The waves of AM radio broadcasts, for example, are about a quarter of a mile long. Those of microwaves range from two feet down to less than 1/12th of an inch.

Microwaves themselves were not new. The first deliberately-made electromagnetic waves, produced by Heinrich Hertz in 1888, were microwaves. However, during the early days of radio the longer wavelengths were favored because it was difficult to generate large power at microwave wavelengths, and because transmission distances at those frequencies were limited. Beginning with the development of the vacuum tube shortly before World War I, radio made a gradual movement up the frequency spectrum to the shorter wavelengths. During World War II, microwave frequencies achieved much of their fame in radar applications.

Microwave development after the war owed its rapid growth primarily to inherent economies. Although coaxial and other transmission systems could provide similar service and performance, microwave equipment had significant economies in maintenance and operation. It could also provide common carrier remote control and telemetering.

This led to the growth of private industrial communications systems, high speed data transmission and closed-circuit television where the cost of other methods of transmission would have been prohibitive.

Besides offering additional frequency space, microwave transmission allowed many stations to operate on the same frequencies without interference. Microwave frequencies were more easily confined to line-of-sight transmission using antennas which directed the signals in narrow beams.

As early as 1951, Collins Radio Company regarded microwave as a new field of expansion. Previous work at the Collins hangar laboratory in Cedar Rapids showed that the Resnatron could produce high-power microwaves. Microwave studies were transferred to Dallas, where Texas Division engineers studied communication needs of the telephone and pipeline industries. After a year of analysis, Collins management decided to design and manufacture some of the first commercially available microwave equipment.

Two preliminary projects were started: (1) a low power line-of-sight microwave system and (2) a 24-channel carrier system. Collins engineers were satisfied with their early results, so microwave development started in earnest late in 1952. By the spring of 1954 the first prototype was in service between Dallas and Irving, Texas.

Collins sold its first microwave system in 1954 to the California Interstate Telephone Company, and the second sale followed when the company furnished a system to A. J. Hodges Industries, a lumber and oil firm in Shreveport, Louisiana. This system was put into operation in 1955 soon after Collins began mass-producing microwave equipment.

That same year, industry took note of the important advantages of microwave in improving production standards. One of the first of these was recorded when Collins provided a 708-mile communication system between Houston, Texas, and Ponca City, Oklahoma, for the Sinclair and Continental Pipeline Companies. Completed in 1956, the system represented the largest battery-powered, privately-owned microwave system in the world. The project was the first high-density microwave system in the petroleum industry. Al Petrasek, Texas Division sales engineer, negotiated the million-dollar contract.

From that point on, Collins achieved several significant milestones in the production of microwave systems. One of its most important programs started in 1956 when the company began preliminary negotiations with the Federal Aviation Administration. The FAA was searching for a more effective method to control and monitor the movement of air traffic, and required radar surveillance coverage throughout the country. Under three contracts totaling about $25 million, Collins Radio provided FAA microwave remoting systems which made up the world's largest communication and radar data handling complex. Forty major air traffic control centers throughout the United States were linked by 95 microwave systems, totaling 631 individual microwave stations spanning more than 16,000 miles.

From its inception, microwave research and development at Collins centered on a basic design philosophy stressing reliability of the equipment. This led to several significant contributions. The most important was pioneering use of a floating

> A variety of microwave radio equipment undergoes final tests in Dallas prior to shipment.

> Deep snow around some microwave relay towers for the Northern Pacific Railway Company meant Collins had to construct two-story buildings to house equipment and provide shelter for communications maintenance teams during storms. 1969 photo.

› Microwave relay towers were often located in forbidding terrain, such as this station for a British Columbia hydroelectric company.

battery-powered plant to eliminate disruption of communications by power failures. This principle was used in some of the earliest Collins systems. Another contribution came in 1955 when the company decided to manufacture a carrier multiplex apparatus to be compatible with international standard carriers. This design resulted in the use of a fully synchronous multiplex system using the Collins mechanical filter.

The most significant step in microwave development on an international basis came when Collins developed a large system in Venezuela. In 1958, officials of Compania Shell de Venezuela were given a demonstration of a Collins tropospheric scatter communications system which had been developed for the military services. The South American oil company officials decided to use Collins equipment to connect an oil refinery at Cardon to an oil field at Concepcion near the large city of Maracaibo. The next step came when the oil company decided to extend communications circuits from Caracas 234 miles into Maracaibo. The effort was designed to establish two important business centers with dependable communication for the first time. Collins engineers were able to connect the two cities by spanning a 7,200-foot ridge with a tropospheric scatter system, and connecting the scatter system to the two cities with microwave links. In May, 1958, the system was turned on, and for the first time 24-hour telephone and teletype service was available between Caracas, Cardon and Maracaibo.

Collins microwave also furnished a high degree of reliability for military applications. Company microwave and carrier systems provided communications for the Atlas Missile Complex at Frances E. Warren Air Force Base at Cheyenne, Wyoming. Collins engineered, furnished and installed equipment for the Pacific Missile Range at Point Mugu, California. Fort Bliss, Texas, one of the world's busiest missile ranges, installed a range safety and communications system using Collins microwave equipment.

One of the first long microwave systems in the United States was constructed for Mid-Valley Pipeline Communication System. This extensive system stretched from Longview, Texas, to Lima, Ohio, and included a master control station at Longview equipped with 26 channel ends of telemetry, supervisory control and alarm, telegraph and voice.

In 1962, Collins provided the microwave equipment for the first statewide educational television network, established in South Carolina. Collins also provided microwave systems and equipment for other large educational television networks. These installations included the Pennsylvania Public Television Network, Kentucky ETV System, Indiana ETV System, Georgia ETV Network, the University of Texas System, the TAGER System in Dallas, and the Louisiana Hospital Video Network.

By the late 1960s, microwave installations completed or in process included systems for railroads, independent and Bell System telephone companies, control of high voltage power conversion and transmission systems, and a regional gas distribution network.

Western Electric emerged as the largest producer of microwave equipment, due to the vast requirements of the Bell Telephone Systems. By the 1960s, Western Electric manufactured virtually all the high density microwave systems needed for multi-channel telephone trunk lines. But Collins Radio succeeded as a major supplier in the medium density microwave market, which included systems capable of handling anywhere from a few dozen to several hundred circuits.

Once again the success of Collins to compete with a much larger corporation lay in its ability to specialize for a particular segment of the market and to provide top quality products.

> Collins continued installing dependable microwave systems where environmental extremes were encountered, such as high temperatures and humidity in South America and the frigid, windswept spine of the Canadian Rockies.

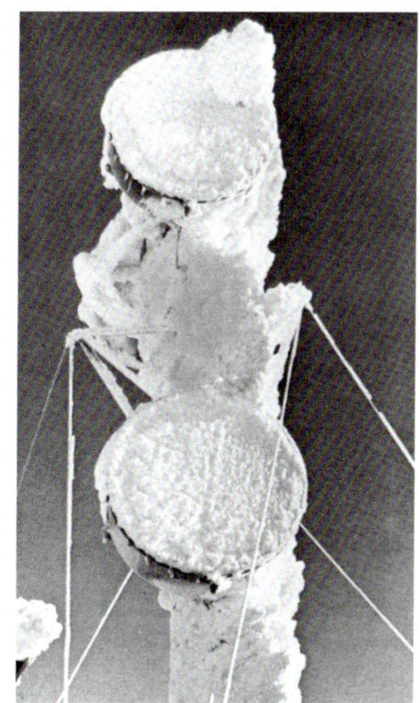

Single sideband

Because of inventions and the application of methods such as the Autotune and Class B modulation, Collins Radio became known for developing techniques and designs, and manufacturing components and subsystems as "building blocks" for a variety of end uses. In turn, these resulted in important new segments of business. One of the most significant for Collins during its history was actually not a market segment, but rather a *method* of communication applied to many markets. It was called single sideband modulation.

Simply stated, single sideband is a form of radio frequency modulation (just as AM and FM are forms) in which the normal carrier signal is eliminated and one of the two modulation sidebands is removed by filtering. Single sideband was not invented at Collins, but the company was the first and principal developer of practical single sideband techniques.

The original development of single sideband came about because of certain limitations in radiotelephone circuits. Experiments were first conducted by John R. Carson of the Bell Research and Development Labs, and the American Telephone & Telegraph Company in 1915. In the 1920s, AT&T used single sideband in regular transatlantic telephone communications. The problem was that it took a whole roomful of equipment to generate and filter a single sideband signal.

In the years following the transatlantic phone service, the use of single sideband was limited to wire and low frequency applications. A general lack of interest in conserving spectrum space led to the opening of other portions of the radio frequency spectrum. In addition, early developments in FM transmission led some to believe that FM might be the ultimate form of voice communications.

The advent of World War II brought an unparalleled need for communication facilities. From this necessity, advances in electronics technology came quickly. Major breakthroughs in basic knowledge and manufacturing techniques during and following the war were important factors in development of HF single sideband communication.

Even back in the 1930, Collins engineers recognized three requirements necessary to make single sideband practical for general communications use: (1) better frequency stability, (2) smaller and lower cost single sideband filters, and (3) better linear amplifiers.

Collins Radio engineers set out to conquer those challenges in the late 1940s. The high frequency portion of the radio spectrum was crowded, and some in the industry thought the congestion threatened to limit the future of high frequency radio. But at

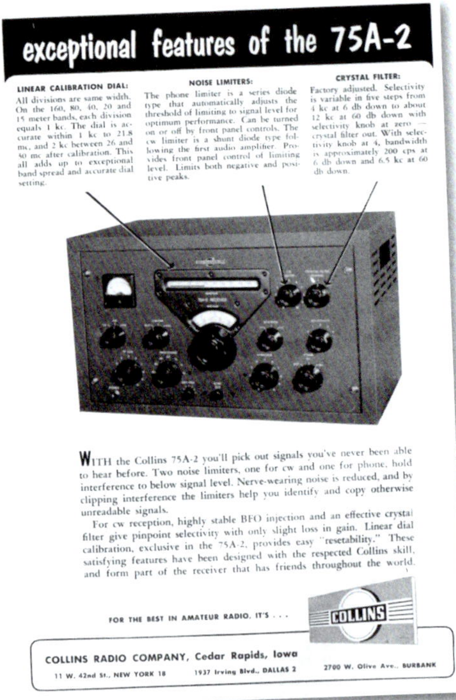

› An original advertisement for the Collins 75A-2 amateur receiver — which originally featured crystal filters — was modifiable to use the newly introduced mechanical filter developed by Collins in 1952.

Collins, researchers realized the importance of high frequencies that could carry around the world. They looked for a way to make high frequency radio more reliable and efficient by rigidly controlling the sidebands radiated by each transmitter to prevent interference. The engineers also sought a way for receivers to select channels much more precisely with less selective fading.

One of the biggest breakthroughs was introduced by Collins in March, 1952. Announcement of a new device called the mechanical filter was made at the Institute of Radio Engineers' 1952 convention in New York. The principle of mechanical filters had been known for many years, but it took Collins engineers to apply them to radio communications. Mechanical filter development started in Cedar Rapids and was transferred to Burbank.

The mechanical filter made possible better control of transmitter sideband radiation and receiver selectivity, so it simplified single sideband communication. The advantages of the small size and increased selectivity of the mechanical filter made it natural for application to common radio problems. The first application was in the Collins 75A-2 amateur receiver. This set was offered with one or more of the filters, depending on the number of selectivity curves the operator wanted.

But these were the Cold War years after the Korean conflict, and most technology advancements were quickly recruited for the military. The United States had bombers in the air on a 24-hour basis, ready for any threat of war. Radio technology had not kept pace with the rapid advances of aviation, and the long-range radios aboard high flying jets often could not communicate with Air Force ground stations.

Armed with the capabilities of the mechanical filter, the previously developed permeability-tuned oscillator and linear tuned circuits a Collins engineering group organized in November, 1952, to investigate single sideband techniques and to develop prototype equipment. The work, under the personal direction of Arthur Collins, took place in the Butler buildings south of the Main Plant in Cedar Rapids. Among the engineers who made significant contributions were Robert Miedke, Vince DeLong, Dale McCoy, William Perkins, Kenneth Rigoni, John Sherwood, Richard Uhrik, Walter Zarris, and Warren Bruene. E. W. Pappenfus was chief engineer for Collins single sideband development activity. The hours of intense study concluded when a course of product development was established.

› Arthur Collins operates single sideband radio gear in a demonstration for Gen. Curtis LeMay. 1956 photo.

In 1952, Arthur Collins predicted 28,000-channel radios would be used on aircraft, because of the new single sideband techniques. The proposed radios for swift aircraft would have power ten times that of the telegraphy radios then in use.

In April, 1953, Robert Mitchell led a group assigned to develop a single sideband exciter and transmitter for the government. That September, Dave Weber headed a group to develop an advanced communications receiver for general commercial use with single sideband techniques.

General Curtis LeMay, commander of the U.S. Air Force Strategic Air Command and himself a ham radio operator, was aware of Collins Radio's progress in single sideband. In 1955, Collins was selected by the Air Force to develop, test, and install a complete single sideband air-ground and point-to-point communication system.

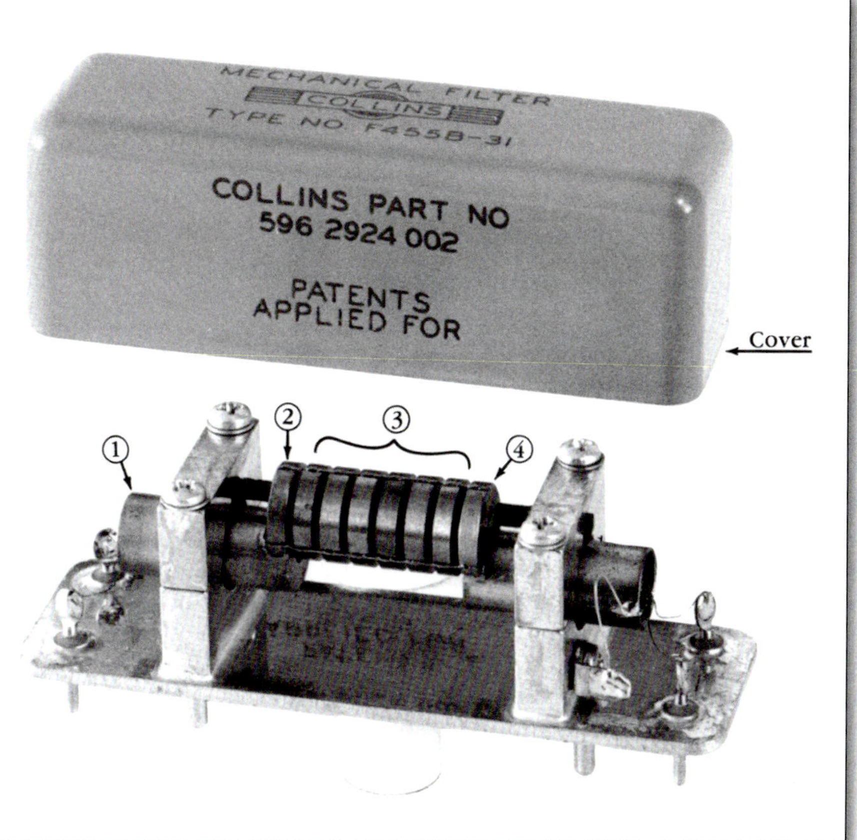

How it works — the mechanical filter

The mechanical filter uses the principle of the tuning fork, taught in high school physics. When struck, the tuning fork sounds a fixed tone in perfect pitch. It also vibrates at its precise tone when a like sound impinges upon it. Collins engineers saw this principle as a way of filtering out different frequency radio signals.

They made a stack of small metal discs, each with a precise resonance frequency, spaced along wires. When electrical impulses are applied to an electromagnet at the input end of the filter (1), the first disc (2) vibrates to the desired signal. Then each disc down the line (3) carries this signal and helps purify the signal by weeding out other vibrations.

The last disc (4) feeds the sorted signal into the main circuit of the receiver. This amounts to a translation of an electrical signal into mechanical force, then translation back into electricity after filtering.

Besides being efficient from an engineer's viewpoint, the mechanical filter performed the job at less cost than other devices could at the time. ▪

> Melvin L. Doelz — who had about 20 patents while at Collins — was responsible for much of the mechanical filter development. Doelz later became vice president and general manager of the Newport Beach facility.

The project was known as "Birdcall." During work on this project, Collins developed techniques in 50-kilowatt antenna switching, steerable beam antennas, and complete remote control of variable equipment functions from a centralized operating point.

In long-range flights to the North Polar region, the Far East, and Europe and Africa during the next two years, the Strategic Air Command, Collins Radio, and amateur radio operators all over the world demonstrated the effectiveness of single sideband in global communication. In flights of an Air Force C97, the airborne station worked all continents and maintained continuous contact with a ham network in the United States.

The equipment — standard Collins single sideband ham gear — was installed in the passenger compartment of the C97 airplane. Operators included General F. H. Griswold, vice commander of the Strategic Air Command; and Arthur Collins, company president.

One of the high points of the experiment was establishing direct interpolar communication between the North and South Poles. Interpolar contact was maintained at regular intervals for three hours one night and for seven hours the following night in July, 1956. Overall results of the flights were outstanding. The airplane maintained virtually 100 percent communication with every sideband-equipped ham contacted.

> Collins unveiled its achievements in single sideband research and development at the 1957 national convention of the Institute of Radio Engineers in New York. The amateur section of the booth (pictured) attracted many visitors to the KWS-1, 75A-4 and the new KWM-1.

Continued on page 83.

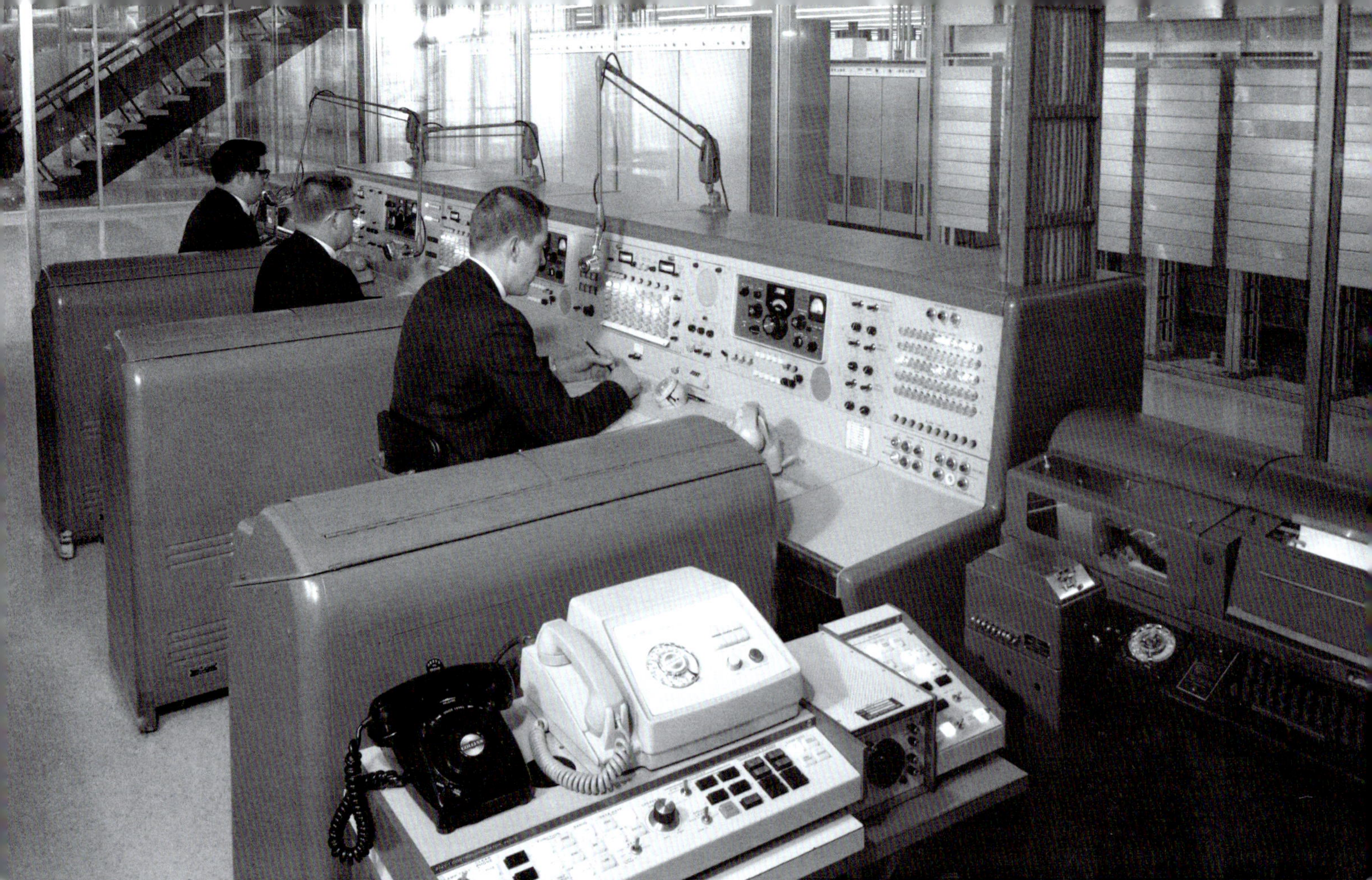

› Comm Central was located in Building 121 in the 1960s. Pictured from left are Bill Ahrens, Ron Wilson and Heinz Blankenhagen. In the foreground is secure voice encryption equipment used with Air Force I communications. Large 45-kilowatt transmitters can be seen through the glass panels.

Comm Central

"COLLINS FLIGHT TEST," "AF3X," "RASPUTIN," "LIBERTY," "KHT," "ROCKWELL FLIGHT TEST," — these past and present call names belong to a unique communications facility operated within the Collins Telecommunications Products Division in Cedar Rapids. Most employees know the station as "Comm Central."

Comm Central was built in 1958 in an area of Building 120 known as Lab 12. The Lab 12 station was used to research high frequency communications in conjunction with the military Short Order Program. A display and observation booth were built around the station to showcase the developments of Collins Radio.

An interim station was built and operated in the Short Order Net from 1960 until the Communications and Data System Division designed and built a complete communications system.

In 1962, the station that many remember as "Liberty" was opened and operated from the new communications and data building (Building 121). The operators called the station the "fish bowl" because of the glass walls.

The Cedar Rapids station became known as "Liberty" during the 1960s when Communications Central was involved with the Andrews VIP network. Collins had a contract with the Air Force to serve as either the primary communication station or as a backup whenever Air Force One, the presidential aircraft, and other aircraft in the VIP fleet carried cabinet members or high-ranking military officers. Over the airwaves the station's call word was "Liberty."

During the Vietnam war years, Comm Central was active in the Military Affiliated Radio System (MARS). During 1967 alone, Comm Central completed 7,122 patches from servicemen in Vietnam to friends and loved ones all across the country.

Comm Central was also involved in many special communication projects. One such project — the Rockwell polar flight, which took place in November, 1965 — was the first around-the-world flight to pass over both the North and South poles. The Boeing 707 established eight world records for jet transports and the crew conducted many scientific experiments. One of the 40 persons on board was Lowell Thomas, Jr. Thomas fed regular voice reports of the flight to Comm Central where they were relayed to Lowell Thomas, Sr., at the CBS studio in New York for his nightly news program. Aboard the aircraft,

Collins provided long-range communication with a 618T-2 transceiver. The purpose of the Collins experiment was to obtain data on high frequency propagation conditions from many locations over a short period of time.

An engineering assistant in Collins' propagation research, John Demuth, was aboard the aircraft to serve as radio operator. At no time throughout the three-day flight were they unable to make contact between the aircraft and the ground station, even from the normally difficult propagation areas surrounding the poles. (*Editor's note: The navigator on the Rockwell polar flight, Loren DeGroot, joined Collins in 1972.*)

Two other 1966 around-the-world business jet flights also were supported by the Liberty station. They were the Learjet flight which established 20 world speed records, and the Rockwell-Standard Aero Commander flight which included Arthur Godfrey as one of its crewmen. A Collins 618T single sideband transceiver was bolted into the luggage rack of the Learjet, and a 26-foot wire was used as an antenna. Over 97 percent communications reliability was achieved around the world on both flights.

An Amelia Earhart commemorative flight by aviator Ann Pelegrino in 1968 was another globe-girdling mission for which the Liberty station provided nearly 100 percent communications reliability.

Plaisted polar expedition

The Plaisted expedition, which used motorized snow sleds to cross the Arctic to the North Pole in 1967 and 1968, also maintained communications with home via Comm Central.

The overland trip to the North Pole started out as a dream conceived by Ralph Plaisted and Dr. Arthur Aufderheide to launch the first surface assault on the North Pole in 58 years.

Although their first attempt in 1967 failed, the team of ten Americans and Canadians tried again the next year and reached its destination on April 19, 1968.

It was with Collins communications equipment that the world learned of the achievement. The equipment was used for communications between the ice party and base camp, base camp and Comm Central, and the ice party and Comm Central. Reports from the ice party dealt with ice conditions, progress made, condition of the men, requests for supply drops, and personal messages.

Plaisted said the Collins transceivers functioned well even though they were subjected to a severe beating over the "most miserable ice anyone could imagine."

Don Powellek, who operated the KWM-2s on the ice and back at base camp, said, "The reliability is way beyond our expectations. We were able to communicate at any time we chose, either to Cedar Rapids or other parts of the world. We have made some fabulous contacts."

First word that Plaisted had reached the pole came through the Collins Comm Central. Charles Kuralt of CBS News was present to talk with Plaisted. The interview was viewed by millions of Americans who watched the Saturday evening news program the following night.

Manhattan project

In the fall of 1969, Collins equipment aboard a huge ice-breaking oil tanker played an important role in a $26 million gamble.

In a dramatic attempt to open the long-sought Northwest Passage through Arctic seas, two United States oil companies and a British firm combined efforts to establish a sea route to rich oil fields discovered at Alaska's Prudhoe Bay. If Arctic oil could be shipped in quantity through the ice-choked waters, the eastern United States would gain access to a vast new fuel supply. The decision by Humble Oil and Refining, Atlantic Richfield and British Petroleum to challenge the Northwest Passage captured the admiration of Arctic experts, but many were skeptical of the chances for the mission's success.

Because of the high priority given to communications and the urgent need to have the system installed, Humble asked Collins Radio to satisfy all the communications requirements of the expedition and put the system in working order within 60 days. The high frequency single sideband equipment installed included a shipboard communications

> To find a new route for shipping oil from Alaska, the *USS Manhattan* broke through frozen Arctic waters. Comm Central handled communications for the voyage in 1969.

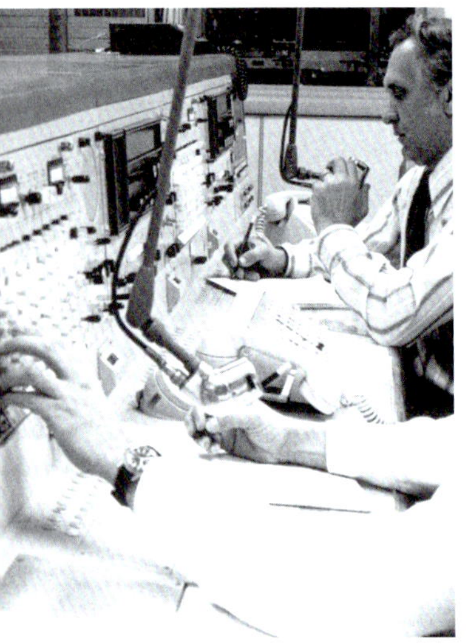

› In the early 1970s, Comm Central moved to a secluded area in Building 121. Radio operator pictured is Joe Dzikonski. In 1983, a new home was constructed for the facility, including a visitor's viewing area.

system, amateur radios, a log periodic antenna and a vertical antenna. The *Manhattan* was the first known commercial vessel to use a log periodic antenna. Tanker-based helicopters were equipped by Collins with homing systems, automatic direction finders and two separate VHF transceivers to allow them to find a hand-held radio beacon used by landing parties. Small VHF radios were supplied by Collins for use by landing parties.

More communications message traffic was handled on the *Manhattan's* three-month voyage than most ships conduct during a lifetime. During its first month at sea, the *Manhattan* averaged 7 hours, 20 minutes of communications traffic per day. The ship made more than 5,000 logged contacts with Comm Central's maritime station, KHT, while at sea, plus additional contacts with commercial carriers handling shortwave radio traffic in the Arctic. About one-fourth of the contacts with KHT were via 100-word-per-minute radioteletype.

Throughout its journey the *Manhattan* maintained 96 percent communications reliability between the ship and the Liberty station, despite such factors as solar flares and magnetic storms. Messages sent from the *Manhattan* were picked up by the Liberty station in Cedar Rapids and relayed to Humble headquarters by way of direct telephone lines.

Flying grandfather

Max Conrad, the famous "Flying Grandfather," departed Winona, Minnesota, on November 30, 1969, on a trip around the world over the poles. This was his second attempt in a twin-engine Piper Aztec named the "St. Louis Woman." Comm Central provided communications for the flight using several calls on high frequency networks. Conrad's aircraft was equipped with a Collins 618T transceiver.

First 747 to London

A single sideband link to the Liberty station in 1970 enabled Americans to witness, through radio network broadcasts, a historic new era of aviation — the first regular service flight of Pan American World Airways' giant Boeing 747 jet from New York to London.

Radio network correspondents aboard the new aircraft gave accounts of the flight. Reports came live via Collins single sideband high frequency radio from the aircraft to the Liberty station, and were fed into landlines to network headquarters in New York.

The 747 inaugural service flight was not the first in which the Liberty station maintained radio contact with a 747 across the Atlantic. When one of the transports was flown from the Boeing plant near Seattle, Washington, non-stop to the 1969 Paris air show, the Collins communications link was utilized for position and progress reports.

The new look

After spending the last ten years of its colorful 25-year history under a stairwell in Building 121, Comm Central was moved and redesigned in 1983 to keep pace with the changing high frequency communications market. Yellow tablets and dial pulse control were replaced by computer terminals and printers. Integrated into the design of the new communications complex was a section dedicated to historical achievements by the various Collins divisions.

The new station features four communication consoles, two mainframe computer systems, a solid-state antenna matrix, and the Collins line of HF-80 communications equipment. A remote-controlled station in Newport Beach is controlled by the station in Cedar Rapids via computer.

Comm Central is licensed as an experimental high frequency research station, an aeronautical flight test station, and as a limited coast maritime station. The radio traffic handled by the station includes communication and phone patching to aircraft, drill rigs, and tankers on the high seas. Comm Central is also involved in marketing demonstrations for customers, and serves as a display area for Collins high frequency products.

Nine operators keep the station manned 24 hours a day, seven days a week. The operators, under the supervision of Heinz Blankenhagen, are: Jack Bond, Lewis Darrah, Justin Dennis, Joe Dzikonski, Ike Hand, Leonard Peters, Rick Plummer, Floyd (Robbie) Robinson, and Ty Smith. ▪

Three B-52Bs of the 93rd Bomb Wing prepare to depart for Castle Air Force Base, California, after their record-setting around-the-world flight in 1957.

Modified Collins 75A-4 receivers were used as the ground stations for the record-setting flight.

In 1957, Collins again demonstrated the advantages of single sideband during the first non-stop, around-the-world flight of three of the Air Force's new B-52 bombers. Modified Collins KWS-1 transmitters and 75A-4 receivers were used as the ground stations on the project.

High frequency single sideband proved itself to the Air Force, providing a spectrum saving of one-half the space ordinarily employed, a nine-to-one improvement in talking power, and immunity to selective fading in the polar regions.

The Short Order Program, a Strategic Air Command communications network project which began in 1958, was based on the operating philosophies which evolved in the Birdcall Project.

"We can now make instant contact with any of our more than 2,000 bombers whether they are at the North Pole or South," said General Griswold in 1960. Short Order, designed and installed by Collins and its systems subsidiary, Alpha Corporation (*see accompanying story*), used single sideband high frequency radio equipment and a vast array of control and switching gear to provide a system that operated automatically under the simple control of an ordinary telephone dial. The system consisted of four stations, all located in the United States. If one of the stations failed to reach a far-reaching aircraft because of propagation difficulties, one or all three other stations were used by remote control via four-wire telephone landlines which connected the stations. Equipment for Short Order was designed, developed and manufactured by the three Collins divisions: Cedar Rapids, Texas and Western, with Alpha Corporation providing system design, management and installation.

But the Collins achievements did more than produce an Air Force contract. Single sideband revolutionized high frequency communications. The company initiated extensive propagation studies to gain more knowledge about single sideband capabilities. Single sideband was soon the standard for civil aviation long-range communication and for a host of other military and commercial applications.

By 1960, single sideband products supplanted aviation electronics as the largest segment of business at Collins Radio, even though single sideband was more accurately a type of communication rather than a classification of business. For example, there was overlapping of the aviation electronics field with that of single sideband, because the transceivers supplied to the Air Force were also used by the overseas airlines.

Three stories underground at the headquarters of the Strategic Air Command, a controller speaks with aircraft over the "Short Order" radio communication system engineered by the Alpha Corporation, systems management subsidiary of Collins Radio. 1961 photo.

Alpha Corporation

By the late 1950s, the size and complexity of electronics installations had increased to such proportions that the most practical way to deal with the problem, as Collins saw it, was through a systems concept. Worldwide systems engineering requirements led to the formation of a new Collins subsidiary — Alpha Corporation.

Incorporated in Texas in 1959, Alpha Corporation extended Collins' activities for the detailed management of space age technical projects, both in the United States and abroad. Headquartered in Richardson, Texas, the company was staffed to design, construct, and install complex government and commercial systems. This included not only each electronic system involved, but also the complete "turnkey" installations with buildings, roads, towers — everything down to the washroom and the lock on the door. After installation, Alpha provided training for customer engineers and technicians assigned to the installation, or furnished complete crews of skilled specialists to staff the finished projects.

Max Burrell, a vice president for the parent company, was made president of Alpha Corporation. John Nyquist was vice president and general manager. Other officers also performed double duty as Alpha and Collins managers.

Although Alpha was established to function as a completely independent entity, program managers drew on the research and manufacturing capabilities of the parent company, and also integrated specialized equipment from other manufacturers into systems when needed.

One of the biggest contracts for Alpha was the Strategic Air Command's "Short Order" communications network. This system consisted of four single sideband ground stations in widely separated sections of the United States. Alpha also did important work for the Pacific Missile Range, and tactical data and multipurpose communications systems for the U.S. Navy.

In 1961, Alpha Corporation was changed from a subsidiary to a division within Collins Radio Company. The following year, Alpha Division and the Texas Division were merged to form the Dallas Division. ▪

› The "billboard" antenna (right) was the first erected north of the engineering building at a site which came to be called the "antenna farm."

Antennas — the ears and voice of radio

Just as the human voice creates pressure waves in the air which the ears detect, antennas play a similar role in radio communication. Antennas are the vital link between signal generating and receiving devices and space, which is the medium that supports radio wave communication. Collins Radio has long recognized the important role the antenna plays in high quality communications systems, and backed this recognition with large investments in equipment and company-sponsored research and development.

In the early days of the company, antenna research personnel numbered so few they did not have a formal group in the Collins organization. Nevertheless, they did have a laboratory — an uninsulated wooden building near the Main Plant in Cedar Rapids during World War II called Fort Dearborn (*See Chapter 3*). Antenna engineers increased in number to comprise two large departments by the mid-1960s. From the beginning, the goal of antenna research at Collins was to provide a high-quality, broad-based

> Hundreds of model log- periodic antennas were constructed in the Collins model shop. 1959 photo.

antenna research and development capability in support and expansion of Collins products. With that goal, the antenna group produced achievements which made it recognized as second to none in the industry.

Aircraft antennas illustrate in the purest sense the successful development of a component in support of a complete airborne electronic system. The "deerhorn" antenna was one such development.

Next was evolution of the first airborne antenna to combine communication and navigation functions in a single unit. The sleek, aerodynamic styling of the 137X-1 earned an award in a design competition in 1959.

Other airborne antennas developed by Collins included antennas for the U.S. Army's AN/ARC-54 airborne FM transceiver. The ARC-54 communication antenna was a 54-inch "whip" mounted on top of a coupler which was remotely operated by the frequency dial on the pilot's communication console.

› Special support towers held and rotated aircraft models for measurements on antennas mounted on the model aircraft. The height provided a free space effect. 1959 photo.

Many high frequency antenna designs were developed at Collins during the 1950s. A basic change was made in high frequency antenna design philosophy to go from narrowband dipole antennas and rhombic antennas to a basic broadband concept, such as the broadband dipole, the discone, the sleeve, and log-periodic antennas. Directional antennas came into wide use, and most were log periodic.

The first antenna built north of the engineering building on the "antenna farm" was called the "billboard." This antenna was made up of seven 120-foot towers arranged in a circle. By selecting a particular radiating tower across from its corresponding reflecting screen, the operator can choose the direction of coverage for his radio signal.

The second antenna installed at the antenna farm was called a logarithmically periodic antenna. The log periodic antenna principle was discovered by Dr. R. H. DuHamel, head of the Collins antenna group in the 1950s. It was called "one of

> This is one of the tropospheric scatter communications systems Collins provided for the U.S. Army. The system operated by scattering radio energy off the troposphere.

the most significant antenna design advancements in recent years." Previously, high frequency radios required several antennas to operate over the full frequency range. The log periodic antenna meant a radio could be tuned through a much larger frequency range without changing antennas. The first log periodic antenna at the antenna farm was later replaced with another having an even greater frequency range.

Another example of how Collins' research resulted in benefits for the company as well as national defense is illustrated by the company's "hard" antenna development and production. A hard antenna is one that is designed to survive effects of nuclear weapons. In 1958, Collins recognized the need for a hard antenna for secure communication sites and in communication between hardened missile complexes. Collins built a retractable telescoping antenna which could be raised to a height of 76 feet. The antenna was used in Civil Defense and similar installations. The experience led to development of antennas for Atlas and Titan hardened intersite communications.

"Hi, I'm from Collins"

A contest for Collins Radio employees which began in the summer of 1956, was held for the last time in the summer of 1957.

In the 1950s, it was company policy to shut down operations for two weeks in the middle of the summer for inventory, so nearly all employees took their vacations at the same time. This created an exodus from Cedar Rapids and other Collins locations, and, incidentally, caused an annual business slump for local retail stores.

Before the annual shutdown in 1956, it was announced that a $25 cash prize would be awarded to the two employees who had a chance meeting at the most distant point from the plant. So employees could spot each other on the highways and in parking lots, all were issued bumper stickers which read, "Hi, I'm from Collins." To verify the meetings, the two employees were required to write a brief account of their meeting in a letter or on a postcard and mail it from the post office nearest their meeting place to the *Collins Column* office in Cedar Rapids.

Everything seemed to go smoothly in 1956. Hundreds of cards were received, and Ken Johnson of Cedar Rapids was declared the winner for his meeting with Doug Johnson of the New York sales office in Caracas, Venezuela. Doug Johnson was disqualified because he was in Caracas on Collins business and not on vacation.

Organizers of the contest were pleased with the results, so they decided to hold it again in 1957. Joe Franey and Leo Voss each were awarded $25 for their chance meeting in North Vancouver, British Columbia.

At first it looked as though the prize money might go to four employees of the Western Division in Burbank. The first cards received from Robert Bryson and Bill Richardson were postmarked Fairbanks, Alaska, and stated that Bryson met Richardson when his helicopter crashed near a gold mine where Richardson was prospecting. The next cards received were from Davos, Switzerland, and claimed that Jim McPherson was operating a cable car at a ski resort when he ran across Bryson and Richardson. Then came notice of a meeting of McPherson and Bob Postal at a Key West, Florida beach, followed by cards postmarked from Nicaragua, where Bryson said he met Richardson as they were both fishing for fresh water sharks in Lake Nicaragua. And finally the meeting to end all meetings — two postcards arrived from Capetown, South Africa, where Richardson said he ran a canoe into the side of an auto ferry, only to be rescued by the ferry captain, who just happened to be Bryson.

When the "world travelers" were investigated by the Western Division industrial relations department, they had only one question, "Does someone have a sense of humor, or are we all fired?"

Apparently someone had a sense of humor, because there was no indication in the *Collins Column* that they were dismissed. ▪

> A "Hi, I'm from Collins" sticker like this was given to employees during 1956 and 1957 as part of a contest in which a $25.00 prize was awarded to the two employees who had a chance meeting at the most distant point from the plant while on their summer vacation.

> The August 1957 issue of *Collins Column* promoted the "Hi, I'm from Collins" summer vacation contest.

In 1959, the antenna research group was called on to design a transportable reflector antenna. Collins was first to develop an inflatable reflector antenna which could survive the outside environment without the protection of an external radome. It was found that inflatable reflectors as large as 15 feet in diameter could be designed for on-sight erection. The techniques developed by Collins have since been followed by many companies in diverse applications.

For scatter communication terminals with smaller power requirements, Collins developed a 10-foot diameter solid surface reflector. Two of the antennas could be unpacked and erected by four men in two hours.

The ability of Collins to meet communication needs with innovative antenna designs played no small role in Collins Radio's entry, and the entry of the United States, into the space age. ▪

UNITED
STATES

The space age

The quarter-century point in the company's history coincided with a significant transition in the operations of the entire electronics industry.

The rapidly growing scope of commercial applications and military and scientific missions to be performed involved increased complexity, uncompromised reliability and exacting performance. From the 1930s through the early 1950s, units such as high frequency transmitters, the Autotune, and the ARC-27 UHF transceiver were individual units developed to meet individual needs of the industry. But by the late 1950s, this traditional approach of product development was becoming obsolete. Particularly in the military field, complete systems which could perform many functions were necessary. These tasks frequently required the cooperative efforts of a group of companies, both large and small, working under the direction of a systems management contractor. The space programs at Collins were typical of such cooperative efforts.

‹ Saturn V and Apollo spacecraft being assembled at Cape Kennedy. Collins supplied complete communications and tracking systems for the Apollo missions.

Early programs in radio astronomy, including use of the moon as a passive relay for communication, development of the radio sextant, and the deep space radio telescope at Anacostia, D.C., gave Collins a strong background in the emerging field of space technology. At the start of the U.S. space program, Collins had perhaps as much knowledge as any company in space communication.

North American Aviation built the first of three X-15 rocket planes in 1957, and Collins Radio received a contract for the communication/navigation system in November of that year. The equipment was delivered in August, 1958, and consisted of elements derived from Collins' standard communications, navigation and identification (CNI) package which it had developed for military jets. Development of the X-15 communication/navigation system was the first venture for Collins, and the United States, in manned space flight.

> Dropped from the wing of a B-52, the X-15 used rocket power to accelerate to speeds of more than 3,600 miles per hour.

A sleek, panatela-shaped craft more like a piloted missile than an aircraft, the X-15 was 50 feet long and had thin wings spanning only 22 feet. It was carried aloft to about 40,000 feet and then launched from its B-52 mothership like a fledgling.

The brief experimental flights in which a test pilot soared to the edge of space made significant contributions to America's aeronautics and space knowledge.

Collins' experience in ground systems began in 1958 when the company installed the first NASA Deep Space Instrumentation Facility at Goldstone, California. Successful completion of this station to provide tracking and communication with a deep space vehicle led to contracts for installation of overseas stations at Woomera, Australia, and Johannesburg, South Africa.

> Between 1958 and 1963, Collins provided a total of 11 space tracking stations. First of these was at the 85-foot dish antenna facility of the Jet Propulsion Laboratory at Goldstone, California, which was used in many of America's deep space satellite probes. Collins supplied all major equipment except the reflector, mount and hydraulic servo mechanism.

Parallel with this effort was the design and installation of a down-range station in Puerto Rico for tracking the initial trajectory of space probes. The overall technical and management capability, which was essential for programs of this magnitude, did not go unnoticed and led to several additional Collins contracts for the Department of Defense and NASA. One of these required that Collins provide two highly versatile tracking and data acquisition systems to the Defense Department for classified programs. These systems tested Collins' ability to do a "turnkey" job. Portions of a satellite tracking facility were provided for the Electronics Proving Ground at Ft. Hauchuca, Arizona, and six data acquisition facilities for the Goddard Space Flight Center were installed by Collins, beginning in 1962.

An early NASA program to advance the study of space communications was Project Echo. To encourage a great deal of scientific research, NASA invited the radio industry at large to participate in the project on a voluntary basis.

Following the announcement of the project, Collins and its systems subsidiary, Alpha Corporation, actively joined with others in the industry to plan maximum use of the Echo satellites for research. Collins/Alpha participants wanted to test tracking techniques to see if satellite links could be used for speech and teletype transmission.

On August 12, 1960, NASA launched Echo I to study the use of passive communication reflectors. When it reached the prescribed altitude, the metal sphere housing the payload separated from the rocket. Bursting like a jack-in-the-box from the divided sphere, hundreds of yards of aluminized Mylar plastic poured out. Inside the uninflated plastic balloon was a residue of ten pounds of powdered benzoic acid. On contact

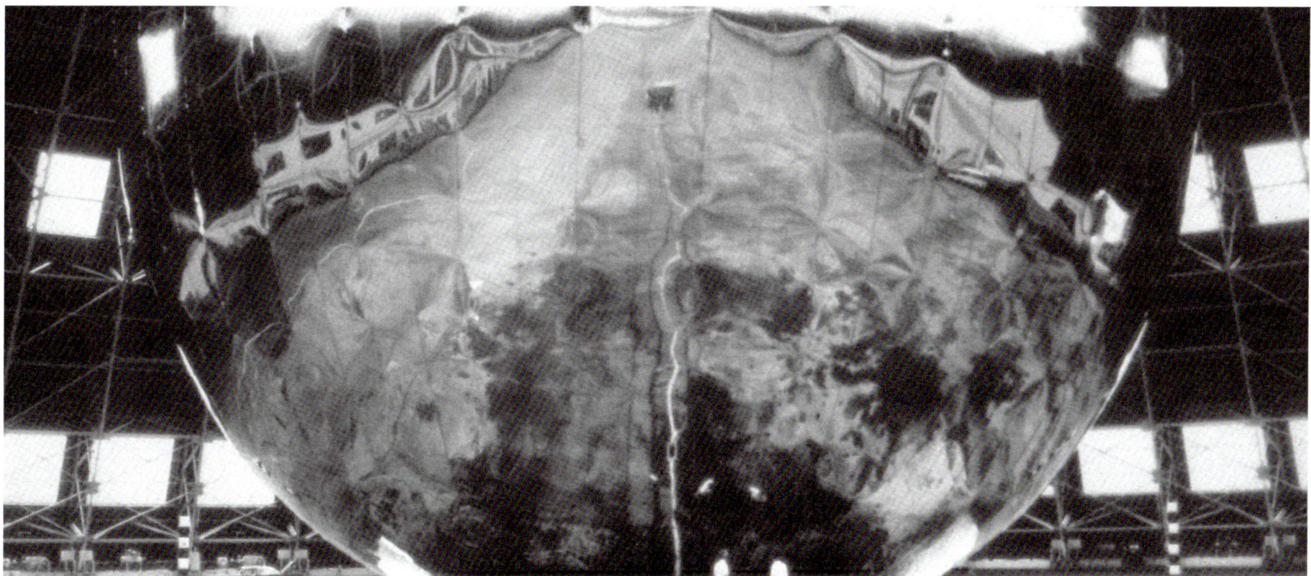

with the thin atmosphere, the limp plastic inflated in a matter of seconds to a firmly rounded, gleaming sphere 100 feet in diameter.

The following day as the satellite rose over the northwest horizon in its eleventh orbit, Collins tracking antennas in Cedar Rapids and Richardson locked on to their target. In Texas, Al Richmond, project engineer for Alpha, spoke into a microphone: "This is KK2XlC in Richardson, calling KA2XDV in Cedar Rapids, Iowa. Do you read me, Cedar Rapids?" The answer came from Cliff Beamer, Collins research scientist: "This is KA2XDV in Cedar Rapids, Iowa, calling KK2XlC in Richardson, Texas. We receive you loud and clear."

The first two-way radio voice transmission via artificial satellite had been accomplished.

The following week, on August 19, 1960, Collins research teams achieved another first. Working with photo transmitting equipment supplied by the Associated Press, a wirephoto of President Eisenhower, taken the previous day in Washington, was transmitted by Echo satellite from Cedar Rapids to Richardson. This was the first photograph transmitted by satellite.

In January, 1962, Echo II was launched but met an untimely end. It inflated too rapidly at an altitude of 200 miles, and the 135-foot balloon exploded into shreds.

The shattered balloon altered, but did not cancel, the plans of Collins engineers and technicians assembled in the company's space tracking station at Cedar Rapids. The Collins crew was able to track and bounce signals from the fragments of the Echo II sphere and received radio signals of "relatively good quality."

The most conspicuous equipment used in the Project Echo experiments were the antennas. Three 28-foot parabolics were installed; one receiving and one transmitting antenna at the tracking station near Cedar Rapids, and the other, a transmitting antenna at the tracking site in Richardson.

Another example of ground tracking experience was Collins' installation of communication, range instrumentation and data handling facilities for the Pacific Missile Range. The range was located throughout the length and breadth of the Pacific Ocean to support firings of tactical, intermediate-range and intercontinental ballistic missiles, and testing of anti-missile missiles, satellites and space reconnaissance vehicles. Collins and Alpha implemented the communication, range instrumentation and data

› Echo I balloon satellite was test inflated on the ground.

NASA photograph.

› The Collins Project Echo tracking station was located north of Cedar Rapids.

› The Collins tracking crew in Cedar Rapids manned the control console during pre-dawn hours as Echo II was launched at Cape Canaveral. Seated from front to rear are Floyd Perkins, Estel Darland and Frank Metecek. Standing are engineers Clifford Beamer (foreground) and Gerald Bergemann.

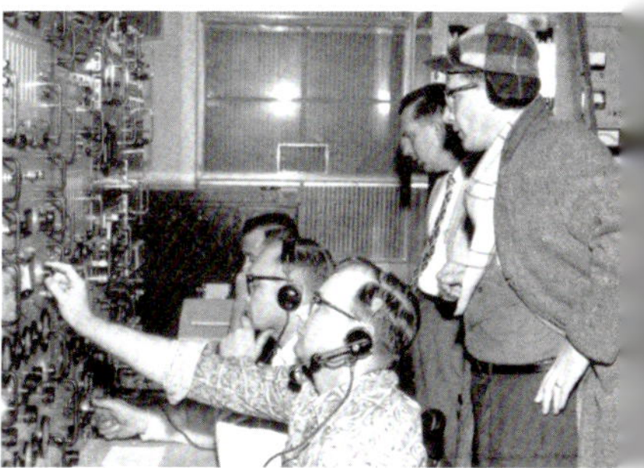

> The first photograph transmitted via satellite was a picture of President Eisenhower, sent from Cedar Rapids to Richardson in 1960.

handling facilities at sites within the continental United States, downrange islands and on range instrumentation ships. High frequency single sideband equipment was furnished by Collins for use both ashore and afloat on range vessels. Every major segment of the Collins organization — Cedar Rapids, Burbank, Dallas and Toronto — made significant contributions to the individual equipment or complete systems for the Pacific Missile Range.

Project Mercury

In 1958, after intense competition with other companies, Collins won the prime contract covering onboard communications equipment for the Mercury manned spaceflight program. As a subcontractor to McDonnell Aircraft Corporation, Collins functioned as a systems manager with sole responsibility for the development, delivery, and satisfactory performance of all subsystem components. The Mercury capsule communication system consisted of 30 components produced during two years of developmental work by Collins and its team of subcontractors. Areas of research included: the functional circuits needed for the several phases of communication; the general problem of signal propagation; several special problem areas in manned-capsule electronics, and each piece of major equipment needed to serve the functions of the capsule system.

The Mercury communications system included several parts:

- *Voice Communication.* The astronaut and ground personnel had to be able to talk to each other during all phases of the mission. Redundant equipment was used at first to provide a backup for both flight and rescue operations.
- *Command Function.* Redundant command receivers with many on-and-off channels were provided to control various functions within the capsule during the launch, flight and reentry.
- *Telemetry.* Two telemetry transmitters relayed scientific, operational and aeromedical data from the capsule to the ground.
- *Precision Tracking.* Two radar transponder beacons in the microwave frequency range were used for precision tracking during flight.
- *Rescue Beacons.* Two rescue beacons, operating on international distress frequencies, helped to determine the capsule's bearings during retrieval operations at sea.

> Astronaut Scott Carpenter talked through a Mercury communications system at Collins in 1960. At left is Eugene Habeger, Collins systems engineer for Project Mercury.

M. Scott Carpenter, one of six astronauts to fly a Mercury mission, visited Collins facilities in Cedar Rapids before the first flight and spoke to employees about their role in the space program.

"Every military person who has had anything to do with an operation of any magnitude is aware of the importance of complete communication. I can't think of any one single factor more important. Loss of communication could mean loss of life and loss of national prestige," Carpenter said.

A full-scale mockup of the Mercury capsule was used to test antenna designs. The Collins engineering team responsible for Project Mercury antennas, from left: Dick Hodges, Paul Zimmerman, Ramsey Decker, Tom Mortimore, Mardis Anderson, George Haines, Jim Shure and Leo Griffee. 1959 photo.

The Mercury communication project team at Collins, headed by Robert Olson and Eugene Habeger, knew that failure of communication electronics in the manned capsule would endanger the astronaut and the mission. The Collins team, along with consultants from Aeronautical Radio, Inc., developed a comprehensive reliability program for the Collins effort, as well as its subcontractors.

The entire first Mercury flight in the spring of 1961 — liftoff to downrange Atlantic return — took just under 16 minutes. The capsule with Captain Alan B. Shepard, Jr., aboard returned safely home. Months of work, study and practice culminated in America's fastest and highest manned flight to date. Shepard's mission was followed by those of Mercury astronauts Grissom, Glenn, Carpenter, Schirra and Cooper.

Collins generated positive publicity during the Mercury missions by boasting that it was one of the first companies to have some of its equipment actually removed from a program. In each of the first five Mercury flights, two Collins sets were carried: a primary system and a backup system. Before the sixth Mercury mission, flown by Gordon Cooper, the backup systems were removed. The record of the previous flights had been so successful that it was deemed safe to rely on one set and use the weight saved to carry other equipment. Collins project team members were elated.

Experience in the one-man-Mercury spacecraft program led to Collins Radio's participation in Gemini — designing and manufacturing the voice communication system which performed successfully on all ten of the two-man Gemini missions.

Collins equipment for the Mercury capsule included, from left: high frequency voice transceiver and backup unit, UHF voice power amplifier, UHF rescue voice transceiver, control panel.

> During Astronaut Edward White's historic space walk, his voice was transmitted to Earth via Collins equipment.

> For the Gemini two-man missions, the Collins voice communications control unit was located directly in front of the astronaut seated on the right (horizontal switches at the bottom of the vertical portion of the panel).

McDonnell photo.

Tragedy caused Apollo design change

Before the last of the Mercury flights, and before the Gemini flights began, plans were launched for the Apollo moon landing program. In December, 1961, it was announced that Collins would supply the complete communications system for the Apollo three-man space missions. The project was described as man's boldest and most extensive program, and one of the most monumental precision jobs in industrial history.

Collins employees designed, developed and manufactured communication and data subsystems used in 22 Apollo command capsules. Most of those modules actually served in space; the rest were used for testing and display.

"Dates were terribly important," said James Westcot, who was Collins' director of business administration on Project Apollo. "North American Aviation (the prime contractor) had an incentive payment program for us if we shipped on time — it was sizable. Overtime didn't mean much. It was getting everything done perfect and shipped on time that counted."

> The deaths of astronauts Grissom, Chaffee and White led to design changes for many Apollo systems, including Collins communications equipment.

In 1964, more than 500 employees working on Project Apollo received a first-hand challenge to provide "defect free" equipment. The challenge came from astronaut Edward H. White II.

"Unless you do your job, I am not going to be able to do my job," White told workers gathered in the Collins cafeteria.

Collins employees who talked with White and Roger Chaffee during their visit to Cedar Rapids were shocked and saddened three years later when a fire fed by pure oxygen flared through an Apollo capsule being tested on its launch pad at Cape Kennedy. White, Virgil Grissom, and Chaffee, the team of astronauts slated to make the first lunar landing, died in the fire. Critics of America's space program emerged in full cry, and all phases of Project Apollo came under intense scrutiny.

The fire and subsequent investigation brought policy changes at NASA, and had a major impact on Collins' portion of the project. Before the accident, NASA specified the spacecraft equipment be designed so astronauts could perform inflight maintenance. For Collins, this meant all communications equipment had to be modular in construction. Therefore, if a fault occurred, the faulty component could be pulled and replaced. Following the fire, NASA abandoned the inflight maintenance policy and asked for a complete redesign from Collins for its portion of the spacecraft equipment. The original design became known as Block I, and Collins' new design was called Block II. Instead of using modular construction, each Block II equipment was a single, solid unit sealed in a special housing.

Development and production of the Apollo project at Collins Radio took nearly five years, and at its peak involved nearly 600 employees. Five other companies were key subcontractors to Collins for various units of the system, with Collins acting as system manager. Arthur Wulfsberg was program director, and Carl Henrici headed the subcontracting organization for Collins. Richard Pickering served as director of systems engineering.

Both individual and group quality performance awards were presented for achieving pre-established goals in the program. Employees earning quality performance awards were eligible for $100 savings bonds to be awarded every six months. "Silver Snoopy" awards were presented to 19 Collins employees, along with letters signed by astronauts expressing appreciation for their outstanding work. Receiving the awards were Jack Stewart, Gerald Hopkins, Lou Christiansen, John Zimmerman, Joe Stoos, Ruth Spurgeon, John Dutton, Boyd Palmer, Vern Jones, Ralph Hepker, Harold Oates, Richard Odell, Richard Rowland, Joe Maerschalk, Dale Thran, Richard Eidemiller, Robert Mitchell, Joseph Dahm and Rodney Peterson.

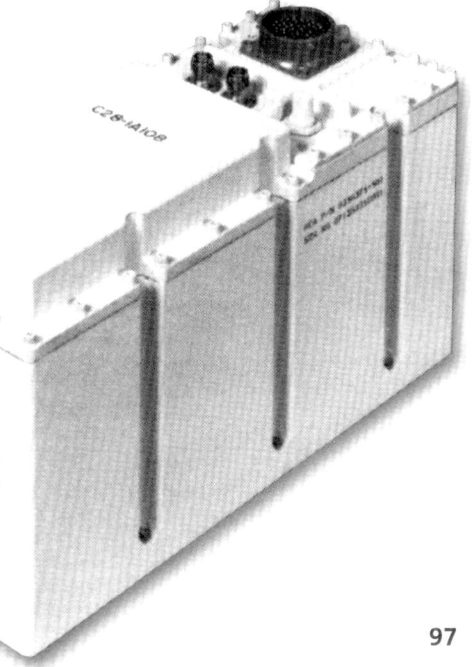

> Block II Apollo communications equipment, such as this VHF/AM radio, was sealed in protective containers.

> Top left: A full-scale mockup of the Apollo spacecraft was built by Collins and put on display for visitors.

> Top right: Assembly and inspection of Apollo equipment was performed in an area in Building 106 in Cedar Rapids. A chart on the wall showed the progress of each type of equipment.

> Twelve stations were used to track and communicate with the Apollo spacecraft as it circled the earth. Each station, including the 30-foot diameter antenna and electronic equipment, was designed, constructed and installed by Collins and a team of subcontractors for NASA.

Communications/tracking system

In 1964, another significant Apollo contract went to Collins when it was selected by NASA to serve as the prime contractor for a worldwide network of communications/tracking earth stations.

The experience gained from Project Echo and the Pacific Missile Range led NASA to select Collins Radio to provide the 15 communications/tracking stations to keep the Apollo spacecraft in contact with mission control. The Unified S-Band System, as the network was known, was an improvement over earlier systems in that a single carrier frequency was used to convey all communications, as well as to determine the spacecraft's position. The $50 million contract was at that time the largest ever awarded to the Dallas Division.

Twelve stations employing 30-foot diameter antennas tracked the astronauts as they circled the earth. For the long distance trek to the moon and back, three large ground station 85-foot diameter antennas were used. In addition to this network of 15 ground stations, Collins supplied communications and tracking equipment for use aboard ships and aircraft.

An additional $2.7 million Apollo contract was awarded to Collins in 1964. Under subcontract from RCA, Collins provided electronic equipment for the lunar excursion module.

> The S-Band System of communications and tracking stations included shipboard installations, such as on the *General H. H. Arnold* instrumentation ship.

Apollo 11 astronaut salutes the American flag, July 1969.

Pictures from the moon

On July 20, 1969, Apollo 11 landed on the moon. That night much of the world's population watched television signals, transmitted by Collins equipment, of Neil Armstrong and Edwin Aldrin, Jr. taking man's first steps on the moon, while Michael Collins (no relation to Arthur) orbited the moon in the command module.

"One of the things you think about is the feeling of accomplishment and personal participation. Everyone in Cedar Rapids certainly should be overjoyed," Arthur Collins said in a 1969 interview for the *Cedar Rapids Gazette*. Shifting any glory from himself, Collins said, "It was the people who did it." Although the system had been thoroughly checked, Collins said he too was surprised and pleased at the clarity of the transmission. "It was certainly very impressive. It represented a remarkable accomplishment in that it is a narrowband video channel, much less bandwidth than normally required for television."

It was later told that Collins had been invited to appear with Walter Cronkite during CBS's television coverage of the first landing. Collins declined, and let the glamor go elsewhere.

When the Apollo 11 astronauts returned, a hero's welcome awaited them around the world. The guest list for a state dinner for the astronauts, hosted by President Nixon, included Mr. and Mrs. Arthur A. Collins. ▪

COLLINS
SSB

Growing markets

The year 1957 was a year of paradox for Collins Radio Company. It produced a greater volume of goods and services than at any time in its history, but net profit declined from the previous year.

It was a year in which Collins introduced many new products, especially in growing commercial fields, but research and development charges and tooling and start-up costs were high in relation to initial sales.

At the end of the 1957 fiscal year, the company's backlog of orders totaled $115 million — 25 percent higher than the year before. Much of the backlog consisted of research and development contracts.

‹ Collins telecommunications equipment was displayed at a 1960 Navy show in Washington, D.C. The company pioneered development and production of vehicular communications systems since World War II, and vehicular radios were used extensively in Vietnam.

› As 1957 came to end, plans were made for a new 210,000-square-foot fabrication facility in Cedar Rapids to help with the companies growth in the commercial markets.

Plans were made for constructing new facilities, financed by the sale of $8 million in debentures. A wholly-owned subsidiary, Texical, was formed to take title of real estate previously acquired by Collins in Cedar Rapids and Dallas, and to assume contracts for construction of new facilities. Under this arrangement, construction began at Dallas on a $1.5 million laboratory building to consolidate all engineering activities of the Texas Division. In Cedar Rapids, a $2.75 million project began for a new fabrication building (now known as Building 105) directly west of the engineering building. Two years earlier, Collins expanded its Cedar Rapids Division by leasing a 21,000-square-foot factory in nearby Anamosa. Elmer Koehn was named manager for the Anamosa plant, which manufactured subassemblies.

Though Collins' long-range product strategy included increasing transition from military to commercial uses, the yearly success of the company continued to rely heavily on government spending. In 1952, 90 percent of sales were to the U.S. Government. By 1957, the percentage of commercial sales had increased, but still stood at only 20 percent.

To broaden its base, Collins made an all-out effort to win new commercial contracts, particularly in the market for communication and navigation equipment for new jet airliners and expanding business aviation fleets.

> The 210,000-square-foot fabrication facility at Cedar Rapids (Building 105) was completed in 1959.

> The plant in Anamosa began operations in 1955.

While it was successful in doing so, the initial costs in terms of engineering, tooling and start-up were large.

Despite attempts at Collins to become less dependent on military contracts, Department of Defense contract rescheduling in late 1957, coupled with the large facilities expansion program, hurt the company's financial picture. Big contracts, such as those for ARC-27 and GRC-27 UHF radios, were coming to a close.

As a result, operating results for the 1958 fiscal year were termed "unsatisfactory" in the annual report. The company showed a net loss of about $250,000 for the year, before a federal tax refund was figured in. Total employment at all locations dropped from 10,000 in 1957 to 7,750 the following year.

"Unsatisfactory though the year has been in terms of the financial results of its operations," Arthur Collins wrote in the company's 1958 annual report, "much has been accomplished in terms of operational reorganization of the parent company into five separate divisions, each of which has been made autonomous as to management. There has also been a modernization of facilities to improve efficiency, maintenance of working capital position and a substantial liquidation of inventories."

One year later, Collins reported its reorganization plans were paying off. Profits for 1959 were the highest in the company's 26-year history, topping $3.7 million. Employment rose from less than 8,000 to nearly 11,000. During the year an important phase of the company's long-range building program was completed with the addition of the new manufacturing building in Cedar Rapids and a new laboratory at Richardson. Plans were announced to build an additional 130,000-square-foot engineering and administrative building in Cedar Rapids (Building 106), and a 120,000-square-foot administrative and laboratory building at Newport Beach, California.

The Newport Beach facility, completed in 1961, housed the Information Science Center, where new designs in data communications equipment were undertaken by Collins Radio. West Coast operations moved from Burbank to Newport Beach in 1961, and the new facility became the headquarters for more than 1,000 employees.

> Production areas in Building 106 featured "teleguides" in front of assembly operators to give detailed assembly instructions via an audio-visual presentation. 1961 photo.

Service division organized

In October, 1960, Collins Radio Company began a comprehensive review of field support requirements for the growing number of Collins equipments and systems in use around the globe. Company personnel experienced in support and maintenance

The Newport Beach, California, facility was completed in 1961.

activities and in marketing research techniques studied the problems of parts provisioning, field service engineering, factory repair and modifications, service parts supply, maintenance documentation and training in the electronics industry.

Out of this study plans were drawn to establish a service operation geared to the same standards of performance and quality as Collins equipment. The service division began operations on Feb. 1, 1961. For the first time in company history, technical service and support responsibilities were consolidated under one management. The division's 625 employees were headed by R. F. Haglund, general manager. Regional field offices were located in New York, Miami, Burbank, and Seattle. International field service support was offered through offices in London, Toronto, Rio de Janeiro, Tokyo, Melbourne, and Frankfurt.

One key growth factor which led to establishing a separate service division was the wider use of integrated electronic systems produced by Collins. The increasing sophistication of such systems multiplied the requirements for competent technical support.

After a year of operation, the division held a contest to find a slogan to use throughout the electronics industry. Leon Glynn submitted the winning entry: "Collins Service: Around the Clock — Around the World."

Avionics

During the 1950s, aviation electronics (or "avionics" as the field was called) was the largest segment of Collins' business. By the end of the decade, the rate of growth had slowed somewhat, but avionics continued to be an important, growing and profitable business area for Collins.

Since its entry to avionics immediately following World War II and until the mid 1950s, Collins concentrated in the commercial aviation market on airliners and heavy business aircraft by providing equipment with the highest standards of performance and reliability. This emphasis paid off. Collins was recognized as a leader in both commercial and military aviation products. More than 80 percent of the new commercial jets — including the Boeing 707, Douglas DC-8, Convair 880, and the Lockheed Electra turboprop — carried Collins communication and navigation equipment. That figure excluded weather radar and autopilots — fields in which Collins was active in the 1950s but still a relative newcomer.

MOS chips were electrically tested by computer-generated test sequences at the Newport Beach facility. Operator is Erna Sanches. Lab technician is Chester Bishop.

Air Force One, the Presidential airplane, was placed in service in 1962 using communications equipment developed and manufactured by Collins. The aircraft is a VC-137, military version of the Boeing 707 airliner, modified to meet the special requirements of flying the President.

Foreseeing the need for lighter equipment to increase the payload of large aircraft as well as to meet the need in small aircraft for professional quality avionics, Collins in 1956 undertook a new research and development program. The goal was to produce flexible communications/navigation/ flight control packages offering savings in size and weight, improve performance and reliability, and provide easier maintenance.

By using newer, lighter components it was possible for engineers to cut weight up to 60 percent. Using transistors where practical reduced size and power requirements. The use of modular design increased the flexibility of the equipment for varied applications, simplified maintenance and improved shock resistance and thermal design.

The first product designed for the expanding light aircraft market was the 17L-8 VHF transmitter. It represented quite a departure from other Collins airborne equipment, because it was designed for mounting behind the cockpit instrument panel. The radio offered commercial airline quality and reliability in a small size.

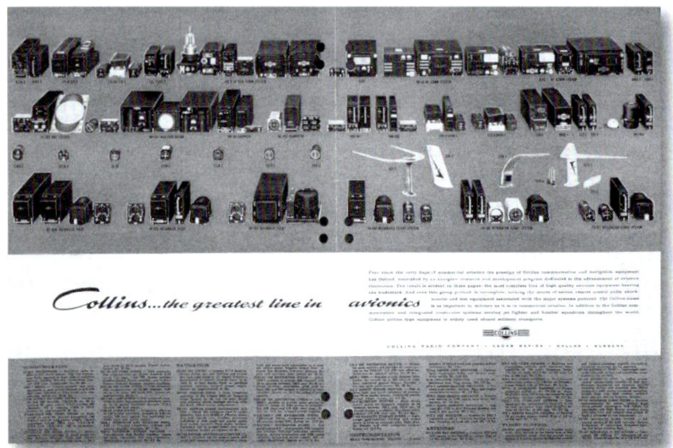

This 1959 advertisement placed in several aviation magazines spoke of "The most complete line of high quality avionics equipment bearing one trademark."

In 1958, Collins introduced its lightweight communication/ navigation packages designed for all types of aircraft. Systems were configured for single-engine, light twin-engine, and medium and heavy twin-engine aircraft. Included in this line was the 618F radio, the first VHF transceiver specifically designed for the general aviation market.

In 1960, the Doppler navigation system was added to the Collins line of avionics. By beaming microwave energy from an airplane toward the earth at several angles, the system, when used with a heading reference and a navigation computer, served as a completely self-contained navigation device.

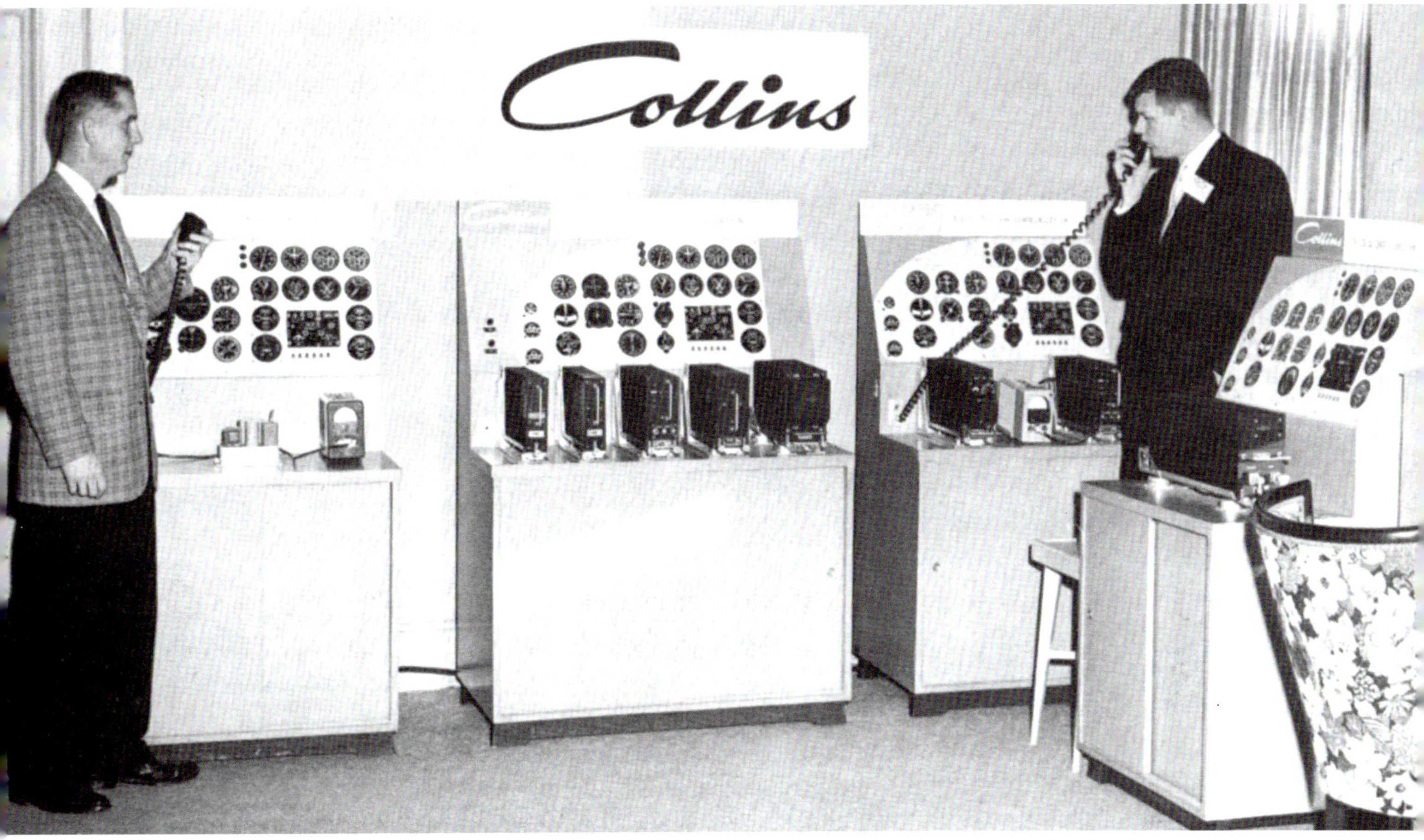

> Lightweight communication/navigation packages were exhibited at the Hotel Adolphus in Dallas in 1958. Pictured are Cal Glade and Craig Christie.

A navigation controversy

A requirement for a new type of airborne radio navigation aid was first generated by the airlines in 1951. The large air carriers were looking ahead to the inauguration of jet aircraft service by 1960. The decision on the first DME (distance measuring equipment), which used a pulse multiplex system and simple electronic ranging circuits, was delayed when the military came out in support of TACAN (tactical air navigation), another type of precise navigation system. TACAN, the military maintained, offered greater bearing accuracy flexibility, and could even be installed on ships. The two systems were incompatible from the standpoints of frequency allocation, economics and air traffic control.

The "TACANtroversy" raged until the President's Air Coordinating Committee established the common VORTAC system in 1956. TACAN facilities were installed at the VOR stations, with civil users relying on VOR and the military on TACAN for bearing guidance, and both relying on TACAN for distance information. A new high-performance DME system compatible with TACAN was developed in the early 1960s.

> The Collins 860E-2 distance measuring equipment featured solid state plug-in boards to make maintenance easier. For dual installation, a single indicator in the cockpit could show the distance to two different ground stations. 1966 photos.

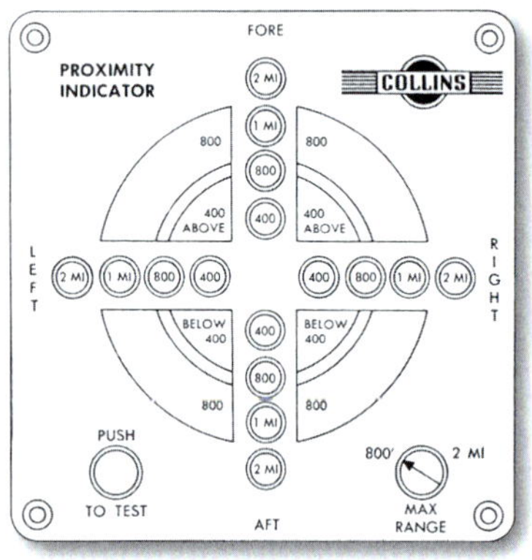

› This drawing shows what Collins' proposed indicator would have looked like for a collision avoidance system in the 1950s. Quadrant and range information were to be shown by four rows of four lights. Upper and lower pairs of pie-shaped sectors would display hemispheric indication, including range.

A system that never got off the ground

Soon after the first flight of the Wright brothers flying machine on December 17, 1903, the aviation industry became acutely aware of the probability of midair collisions. In training and combat during World War I and World War II, both fledgling and seasoned aviators met untimely demises when their aircraft collided in that seemingly vast sea of blue sky.

With the tremendous proliferation of air commerce in the 1950s, it was only a matter of time before a catastrophic midair collision occurred in U.S. airspace. The collision of a Douglas DC-7 and Lockheed Constellation over the Grand Canyon in 1956 became the catalyst for investigation of possible collision avoidance systems.

The Air Transport Association began an earnest quest for a developmental program on collision avoidance equipment. From several proposals submitted to the Air Transport Association, Collins was the first company selected to design and test a "noncooperative" collision avoidance system. Noncooperative meant that the equipment could detect all other aircraft, regardless of whether the others also had collision avoidance gear.

The basic requirement of a noncooperative system was a proximity indicator (PWI) that could determine the direction and range of other aircraft in the immediate vicinity. The effectiveness of such a system was very limited because the PWI required the flight crew to make visual contact with the intruder, executing evasive maneuvers at high closure rates. Additionally, an antenna of approximately 12 feet in diameter would have to be mounted on aircraft participating in the collision avoidance program.

Prototype systems were flown in company aircraft before it was conceded that a noncooperative system was not feasible with the electronic technology of the 1950s.

Competing airframe and avionics manufacturers tried unsuccessfully to gain FAA approval on varied collision avoidance systems. RCA submitted the SECANT system, McDonnell Douglas developed a time/frequency system and Honeywell produced the AVOIDS system.

In 1963, an industry-wide effort was undertaken to develop a "cooperative" collision avoidance system. To be considered a cooperative CAS, each aircraft had to be operating its equipment with the same time synchronization clock. Without accurate timing, collision avoidance computations could not be legitimate.

Having made very limited progress with collision avoidance systems during 1961 and 1962, the Federal Aviation Administration sought technical support from Collins. An effort was made to develop three cooperative CASs.

Through experimentation and practical application, it was apparent that the Time Frequency Range Altitude method of collision avoidance offered the most reliable means of detecting intruding aircraft.

Using data from Collins Radio, the Air Transport Association formed a CAS subcommittee in 1963. Comprised of leading manufacturers in the aircraft and avionics industry, the "Technical Working Group" determined that a noncooperative CAS was again not feasible with state-of-the-art avionics and engineering technology. Thousands of man-hours were spent in closed sessions during 1966 and 1967 to discuss the probable success of a cooperative Time Frequency Range Altitude collision avoidance system. By 1968, aircraft simulation programs were being designed for U.S. air carriers. A limited number of the (TFRA) CAS system simulators were delivered to the Piedmont Airline Pilot Training Center for evaluation.

Although the (TFRA) CAS system appeared to present a viable alternative for interrogating and avoiding intruder aircraft, this system required that all participating aircraft have compatible airborne collision avoidance systems. Considering that inflight aircraft testing was yielding poor results in the terminal area, and equipment costs would be exorbitantly high, it was not surprising the airlines were opposed to the Air Transport Association's recommendation that they purchase collision avoidance systems. Additionally, the general aviation version of the Time Frequency Range Altitude collision avoidance system was far too costly to manufacture to the complexity of the equipment.

Collins has not actively pursued the collision avoidance issue since 1969 because the Federal Aviation Administration and aviation industry at large have not made a firm decision about the direction which should be followed. Three distinct approaches, the Discreet Address Beacon System (DABS), Beacon Collision Avoidance System (BCAS) and Threat/Alert Collision Avoidance System (TCAS) have been reviewed, with TCAS favored by the FAA in 1983. ▪

› A tour of Collins facilities was quickly arranged one day in 1962 when entertainer Danny Kaye flew his private airplane to Cedar Rapids for a visit. "My plane has much Collins equipment, and Collins makes the finest radio equipment in the world," Kaye told a reporter. Pictured from left are Kaye, Bob Winston, John Dayhoff, Glenn Bergmann and Clayton Lander.

By 1966, more than 800 new DME-equipped VOR stations were in operation in the United States, and distance service was also available from more than 200 military TACAN stations. Collins accommodated by building DME units for commercial aviation and TACAN units for the military.

The DME airborne unit exchanged coded pulses with the ground station. The receiver in the aircraft measured the elapsed time between the transmitted pulses and the received pulses, and computed the aircraft's distance from the station. With DME, pilots could give more accurate position reports and estimated times of arrival. In turn, air traffic control could establish more accurate holding patterns and greater safety.

Introductions of Doppler navigation, transponders, and DME were closely followed in the fall of 1961 by what Paul Wulfsberg, Collins development department head, called "the biggest step in terms of technical advancement we have ever made between one generation of airborne equipment and another." Wulfsberg was referring to Collins' new Solid State Systems line. New developments in transistors, diodes and other solid state components allowed their use in the demanding environments of avionics. In many cases the new solid state design did away with all moving parts such as switches, gears, motors, and relays within the airborne units. Production of the solid state systems was designed to meet schedules of the second generation of jet airliners, such as the Boeing 727 in 1963.

The third dimension

Webster defines parallax as "the apparent displacement of an object observed due to a change or difference in the position of the observer." A common example of this, as it pertains to instrument interpretation, is the automobile speedometer which indicates a different speed to the passenger in the right front seat than to the driver who is seated directly in front of the instrument.

In instruments where exact readings are critical, parallax is reduced to an absolute minimum by positioning the pointer as closely as possible to the instrument face.

› Single sideband radio equipment for U.S. Army military vehicles was another growing market of the 1960s. To demonstrate the variety of Collins equipment for a visiting general, jeeps equipped with Collins radios were lined up in front of Building 121. Picture from left: Arlo Meyer, Art Kemper, Paul Hoffman, Clay Kimsey, Leon Griswold, Joe Rosenberger, Bob Cribbs, Karl Stanley, Ed Andrade, Marv Gehr, Marion Albaugh, Roger Herreid, Ray Ruggiero, Evan Maloney, Clellan Wildes, Will Ogle, Hank Dmitruk, Bill Crawford, Jim Adelson, Wes Steele and Don Kent. 1961 photo.

This is simple enough, but what if more than one pointer is required and each requires full freedom of movement? What could be done if the background of the instrument must move and six, eight or even ten indicators must also move and retain their proper relationship regardless of the angle from which the instrument is viewed?

This was the problem handed to Collins engineers in 1961. The instrument specifically under consideration was a new flight director indicator which would enable jet airliners to operate effectively under the same landing minimums as conventionally-powered aircraft.

It was a real challenge, and Collins engineers, principally Harry Passman and Horst Schweighofer, decided on a radical new approach. They changed the way the information was displayed by consolidating pitch and roll commands into a V-shaped bar, and changing the aircraft symbol to a large delta shape. The idea was simple and elegant. It was the natural way to display flight information, as pilots and other flight instrument manufacturers soon discovered.

The first FD-108s incorporating the three-dimensional V-bar steering were delivered to Trans World Airlines in 1963 for new Boeing 727 jetliners. Braniff Airways installed V-bar indicators in new BAC-111s and retrofitted them in Boeing 707s and 720s. Eventually the V-bar became an industry standard, and most avionics manufacturers now offer a choice of V-bar or cross-pointer steering on their flight director instruments.

Down the path to zero-zero

Since its entry into avionics, Collins was a major contributor to battle against aviation's chief enemy — bad weather.

In 1964, the Federal Aviation Administration estimated weather delays, diversions, and cancellations cost the airlines more than $67 million that year. The industry wanted a solution to the costly problem. Working with the airlines, the pilots, and the avionics manufacturers, the FAA established a schedule for achieving "zero-zero" landings

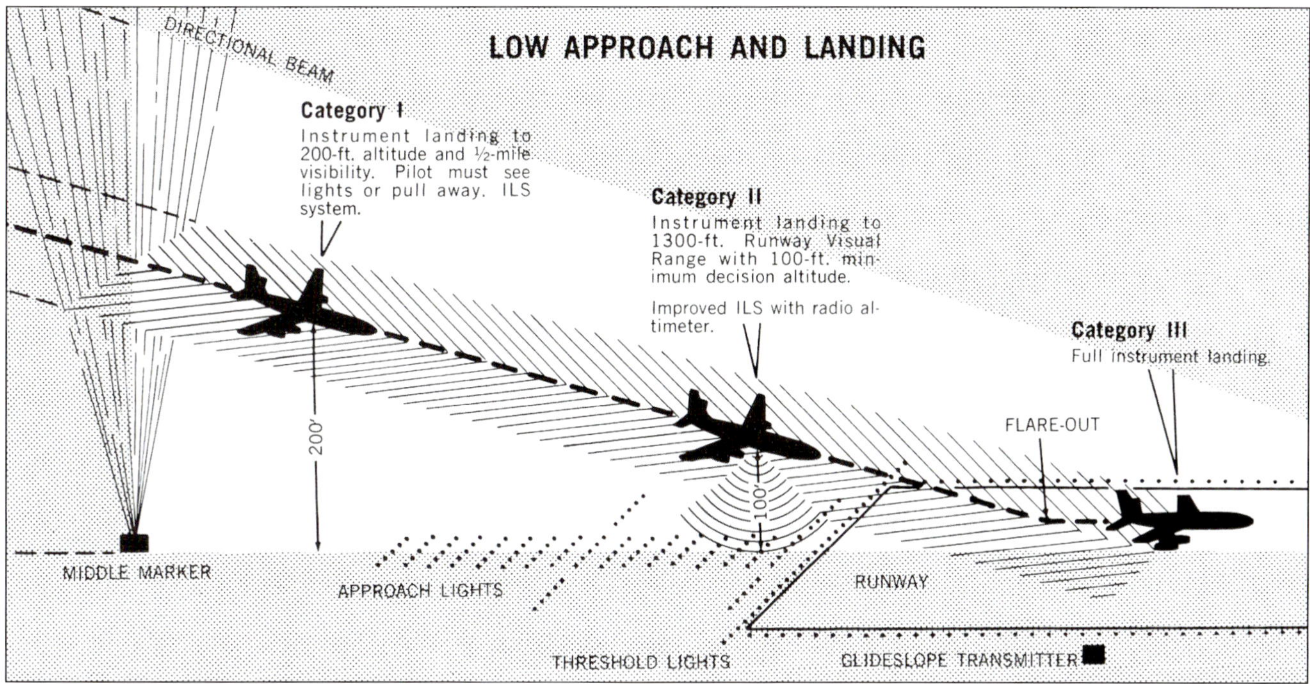

> Chart depicts step-by-step categories involved in achieving all-weather landing capability.

(landings accomplished with a zero altitude cloud ceiling and zero visibility).

The first step, called Category I, designated that airplanes were legally allowed to land with a minimum of a 200-foot ceiling over the airport and one-half-mile of visibility at the runway threshold. These were the legal minimums at the time the categories were established.

Category 11, the next step, meant that certified airplanes could land with a 100-foot ceiling and one-quarter-mile runway visibility. Collins developed the avionics which enabled the first certified Category II landing, achieved on January 12, 1966, by a Jet Commander business airplane.

Next came Category IIIA, which brought the minimum runway visibility requirement down to 700 feet. Again, Collins was the first avionics manufacturer to have its equipment certified. The effort began in 1968 when a team of Collins Radio and

> Representatives from Collins and Lear Siegler display some of the equipment for the L-1011 flight control program. From left: Jack Burns, president of Lear Siegler; Robert Dunn, vice president and general manager of Collins Radio in Cedar Rapids; Jo DeHaven, secretary for Lockheed-California; Harry Kohl, vice president of Lockheed-California in charge of L-1011 production; and J. H. Hanshue, director of material, Lockheed-California.

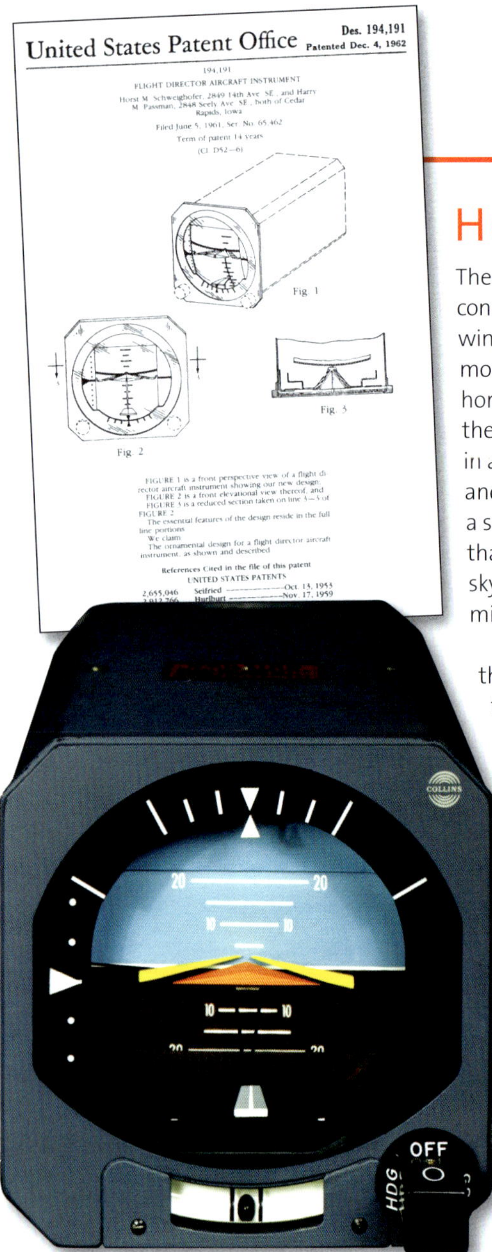

United States Patent Office — Des. 194,191 — Patented Dec. 4, 1962

194,191

FLIGHT DIRECTOR AIRCRAFT INSTRUMENT

Horst M. Schweighofer, 2849 14th Ave. SE., and Harry M. Passman, 2848 Seely Ave. SE., both of Cedar Rapids, Iowa

Filed June 5, 1961, Ser. No. 65,462

Term of patent 14 years

(Cl. D52—6)

FIGURE 1 is a front perspective view of a flight director aircraft instrument showing our new design;
FIGURE 2 is a front elevational view thereof, and
FIGURE 3 is a reduced section taken on line 3—3 of FIGURE 2.
The essential features of the design reside in the full line portions.
We claim:
The ornamental design for a flight director aircraft instrument, as shown and described.

References Cited in the file of this patent

UNITED STATES PATENTS

2,655,046	Seifried	Oct. 13, 1953
[illegible]	Hurlburt	Nov. 17, 1959

› The "V-bar" indicator gave pilots a new way to view flight information.

How it works — the V-bar indicator

The Collins V-bar flight director indicator consists of a small orange-colored, delta-winged airplane symbol appropriately mounted in front of a two-color movable horizon background. This indicates the aircraft's pitch and roll attitude in a conventional relationship. Pitch and roll commands are provided by a set of yellow-colored V-shaped bars that create the illusion of a track in the sky which the pilot follows with the miniature aircraft.

In an approach to a runway with the instrument, if the airplane needs to fly up and to the right to intercept the glideslope and localizer signals from the ground, the V-bars will be positioned above the airplane and tilted to the right.

Pilot reactions to commands presented this way are instinctive and positive. Because both pitch and roll commands are presented on one indicator, the pilot makes simultaneous corrections of pitch and bank rather than making them separately as is often the case when these commands are presented on separate needles.

Commands presented on the V-bars can be derived from a variety of sources. In the pitch axis, for example, the V-bars may be used to command a selected climbout angle, maintain a selected altitude or follow the glideslope on an instrument approach. In the roll axis, the V-bars can be used to command bank angles required to reach and maintain a magnetic heading or follow a VOR or localizer course.

In addition to steering commands, the pilot of an aircraft making an instrument approach wants to see his actual aircraft position relative to the localizer and glidesloper to make sure the steering computer is working properly. In the Collins 3-D indicator, the aircraft position relative to the glideslope is shown by a pointer and scale on the left side of the instrument. Displacement from the localizer centerline is shown by the lateral movement of a miniature runway symbol at the bottom of the instrument. ▪

Lear Siegler was selected by Lockheed Aircraft to develop the flight control system for its new L-1011 TriStar jetliner. The pairing of Collins with Lear Siegler of Santa Monica, California, combined Lear Siegler's experience in automatic landing flareout techniques with Collins' record for airline service and support, derived from other avionics business with the carriers.

The contract was one of the largest ever awarded in commercial avionics history, covering 350 aircraft and amounting to more than $40 million.

On April 27, 1971, the Collins FCS-110 automatic flight control system was the first equipment of its kind to be certified by the FAA for Category IIIA landings.

"It was phenomenal for such a sophisticated system to perform so perfectly on the first try," said E. L. Joiner, the TriStar's chief flight test engineer for Lockheed. Joiner said the pilot just pushed a button "and the plane did the rest."

An example of the efficiency of the automatic system was demonstrated in a subsequent landing at Palmdale, California, site of Lockheed's test facility. A conventional commercial jet airliner making a manual approach under gusty crosswind conditions was rolling and pitching in the turbulence, with the pilot working hard to maintain the desired flight path. The L-1011 following this airliner was using its new autopilot to make an automatic landing. To ground observers, the L-1011 appeared to be "riding on rails" compared to the other airliner.

› An L-1011 makes an automatic landing.

› Pilot and copilot make a "hands off" automatic landing of a Lockheed L-1011 jetliner.

› The AINS-70 automated both short- and long-range navigation for DC-10 jetliners, beginning in the early 1970s. The flight crew could program navigation information on the keyboards and see information displayed on the CRT screens.

In 1970, a new kind of navigation system that made its debut with the new generation of wide-body airliners had far-reaching effects on the way airlines went about their business.

The first system of this new class to go into regular service was called the Collins AINS-70 Area-Inertial Navigation System. Employing a digital computer in its most expanded application in civil air transportation, the AINS-70 automated both short- and long-range navigation, with automatic flight plan execution. It was the first commercial airborne application of the Collins C-System — a digital computer-based system integrating communication, computation and control. Two cathode ray tube (CRT) display units presented each pilot with selected navigation information, and each had a keyboard for the pilot to enter instructions, call for a display of flight data, insert information or modify flight plans. The advanced system was installed aboard the McDonnell Douglas DC-10. ▪

Computer systems

At the beginning of the 1960s, Collins policy-makers felt that an extraordinary increase in industrial data communications was about to take place. Computers had reached the point of development where communication of their data had become a major problem. A single computer had a communications requirement equivalent to the work capacity of many human beings. Collins felt the demand was pressing for new communications systems, and particularly efficient use of the present ones.

Computer design was not new territory for the company. Collins experience in computer systems dated back to research in small airborne units for flight control, and high capacity data transmission in the late 1940s.

‹ By adding or replacing modules of the Collins C-8500 computer series, the user could build and expand the system to meet his requirements.

Practical theories and mathematical analyses of data transmission were formally presented in a 1949 paper, "Radiotelegraphy," by Walter Wirkler, and in a 1951 paper, "Communications Systems Analyses," by M. L. Doelz.

The first results of the research in high-speed data transmission were Predicted Wave Signaling systems, which were based upon improved frequency stability, integration and synchronization techniques. Their capacity to handle information in bits per second was more than twice as efficient as any other form of data signaling developed by that time.

One of the early tests involved a Cedar Rapids-to-Dallas radio circuit in 1952 and 1953. The predicted wave equipment was tested by Collins and Bell Telephone Laboratories in 1954, resulting in selection of predicted wave signaling for the DEW Line.

Data transmission was refined and in 1957, the name was changed to Kineplex.® With its related terminal equipment for handling punch cards, paper, magnetic tape or other information sources, the system constituted Collins' entry into the data communications field. Kineplex, which was fully transistorized, was designed to operate on available communications channels, whether they were telephone lines, microwave, single sideband, or scatter radio transmission.

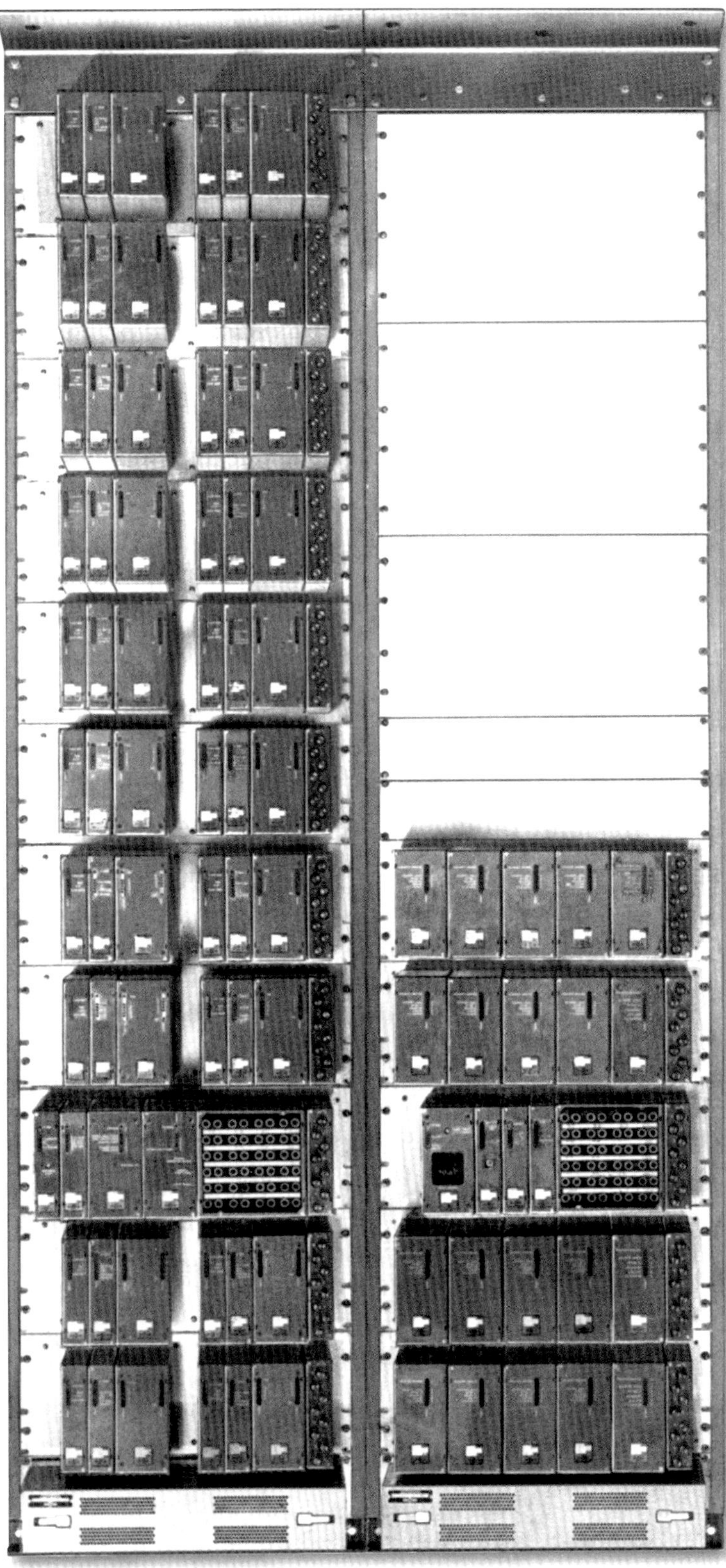

> The Collins TE-202 modem, which was about the size of a refrigerator, was part of the Kineplex system. 1959 photo.

In the late 1950s, the principal customer was the military, which used Kineplex data links to couple processing centers handling missile data, for missile range installations, for tactical deployment, and for worldwide logistics. In one military system, known as Naval Tactical Data System (NTDS), functions such as detecting and tracking enemy and friendly forces were worked out in computers to coordinate an entire naval task force. NTDS gave the task force commander a complete, overall tactical picture of the task force situation.

From the early days of the company's research efforts in the field of synchronous communications, Collins sold electronic devices, called modems, which allowed computers to talk with each other over telephone lines and reliable radio circuits. The first Collins modem was introduced in 1955. The TE-202 was about the size of a refrigerator and weighed 700 pounds. Collins became the world's largest independent producer of data modems at 2,400, 3,600 and 4,800 bits per second. The modems were used in surface and airborne radios, and wireline data systems.

By 1961, Collins Radio had established profitable operations in military and commercial electronics, and its telecommunications business was growing rapidly.

The logical choice for a new segment of business, as Arthur Collins saw it, was digital computers which could handle both data processing and telecommunication message switching, among other tasks. The major competitor was IBM, which dominated wide portions of the computer market at the time. Other companies had spent large sums of money in apparent efforts to parallel the position of IBM, but Collins decided on a different approach.

Collins had been successful in challenging established competition by making an energetic effort toward selected markets. A "rifle rather than a shotgun approach" was how the 1963 annual report described Collins' approach to the computer market. But Arthur Collins felt it could not be a limited effort. If his company failed to develop a complete and superior system, it would be inconsistent with company policy.

Research began for computers that would be true general purpose machines with enough versatility to be equally at home with communication processes (such as teletype message switching, priority routing, and data transmission and conversion)

> The Collins Data Central building (Building 121) in Cedar Rapids was constructed to showcase the company's developments in computer systems and provide a test facility for C-System technology.

as well as with conventional data processing operations. Collins engineers reasoned that communications could borrow much from computer technology, since storing, selecting and processing data were fundamental to both. They recognized that by properly programming a computer, various basic communications processes could be regulated with greater efficiency and accuracy, and with smaller-sized equipment.

Before development work got too far along, Collins engineers looked into existing computers to determine whether they could be adapted or modified for communications work. They found most were inefficient for communications, largely because they could handle a given number of specific problems, but lacked the "general purpose" flexibility a communications-oriented computer needed.

The Collins approach to the problem was to create a flexible computer. A magnetic core memory, which could operate 15 times faster than the main memory, took the place of "fixed-wire" logic.

A second major design feature was to improve both internal and external communication capability. This was done by using a high-speed transfer link or common data bus. In effect, the transfer link, coupled with appropriate data transmission terminals, provided complete freedom of data transfer between all peripheral equipment.

To demonstrate what high speed data transmission could do, Collins established a data processing center in Cedar Rapids in the spring of 1961. The $1 million, two-story structure (Building 121), located next to the engineering building, became the nerve center of the company. Facilities at Texas, Toronto and the new Information Science Center at Newport Beach (*See Chapter 8*) were linked via Kineplex with Collins Data Central. Links were also added to computers operating in Kansas City, Washington and New York.

In addition to serving as a working demonstration of Kineplex, the data processing center was a pioneering effort toward improving basic management techniques through rapid and precise transmission and processing of all types of business data. Inventory, shipping, receiving, production and payroll records and other information necessary for daily operations were processed at the center for the various locations on the system. The centralized data processing concept was expanded to include voice, teletype and facsimile service. As it took shape, it became a communication *and* data processing system, with equal emphasis on both elements.

> Because of the complexity of the C-8400 system, assembly was no easy task, as evidenced by the web of wires which had to be correctly attached.

> Providing a system-oriented configuration and a simplified maintenance program through unit replacement, the C-8500 computer used modules of the same size used by the air transport industry for avionics racking.

> The magnetic core memory of the C-8400 system took the place of "fixed-wire" logic. Because command sequences were stored magnetically in this memory, the Collins computer could be "rewired" for a new application by reading in a new logic program.

After the "bugs" were worked out with the system installed at Collins, a commercial version was developed. Development of the C-8400 System started in May, 1961, and just 13 months later the first pilot system was operating.

One of the first installations of the system was made for Aeronautical Radio, Inc., (ARINC) for use as an airline teletype message processing and switching center at Elk Grove Village, Illinois. The facility interconnected circuits for airline offices throughout the United States and overseas. An installation ordered by the New York Central Railroad represented the first railroad application of such equipment for automatic assembly and distribution of messages and data information. Collins Data Central in Cedar Rapids acted as a classification and distribution point for all the messages transmitted throughout the systems.

An evaluation of the computer systems two years later showed costs related to initial customer orders were running higher than the revenue generated.

"It is anticipated that continuing commitments for servicing these customers will adversely affect 1965 earnings to a lesser extent, and that orders undertaken subsequently will be profitable," Arthur Collins wrote for the 1964 annual report.

In 1966, Collins took another technological step by introducing the C-8500 computer system, which combined multispeed communication with business and scientific data computation for control of on-line, real-time operations. Using microminiature integrated circuit components, the new computer group was designed in modular packages similar to the black boxes manufactured for commercial airliners. This design allowed the customer to customize his system by simply adding or replacing modules.

The C-8500, part of the "C-System," made available to users the first completely integrated communication, computation and control system having virtually unlimited expansion capability.

Arthur Collins, believing the market potential for C-System applications was huge, personally supervised much of the computer system's development and plans for marketing.

"This continuing development of communication, computation and control systems is of particular significance for long-range expansion of market potential," he stated in the 1968 annual report.

With the C-System, it was no longer necessary to isolate accounting, engineering, manufacturing, communications, personnel and general management functions. The C-System had the ability to accommodate all these functions.

In 1966, Collins undertook an extensive program to construct facilities for process production which resulted in addition of 1.4 million square feet, bringing the total facility area to 4.1 million square feet. Also, extensive modifications were made in existing plant space for advanced technical operations. These expansions occurred in Cedar Rapids, Dallas, Newport Beach and Toronto, as well as in Frankfurt, West Germany, and Melbourne, Australia.

The 240,000-square-foot solid state device plant at Newport Beach, placed in operation in 1969, was the final unit of this construction project.

› Collins Data Central in Cedar Rapids after installation of C-8500 equipment.

Collins called this new form of computer-controlled operation "teleproduction," a term which it applied to using sophisticated computers not only in management and design, but also in the output of products or physical work.

Collins provided teleproduction services, the highly automated design, fabrication, and testing facilities for multilayer circuit boards, hybrid microcircuits, and metal oxide semiconductor (MOS) arrays for its own engineering force and outside customers. By the late 1960s, Collins was a major supplier of hybrid microcircuits.

Although multilayer circuit boards had been used for several years, the advancements of MOS/LSI (large scale integration) technology increased the need for dense multilayer interconnects. Collins' use of computer-oriented layout and numerically-controlled fabricating machines provided the capability to manufacture boards of two to 14 layers in as little as 14 days.

In September, 1968, company officials announced the beginning of a vigorous program to sell its C-System to industry and government. Collins believed that it could compete in the computer market and capture a sizeable portion with its system. The company cited several reasons, including easier maintenance, greater flexibility, higher efficiency, and the capability for computer control of manufacturing.

› Each photoplate copy of a multilayer circuit board layer shows the patterns of the various circuits.

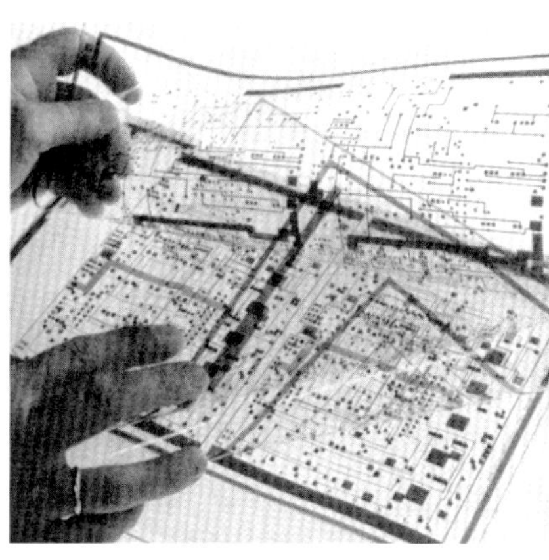

> C-System techniques were applied to teleproduction — using the computer to produce manufactured products. Blank multilayer circuit boards were fed by a special conveyor into machines that etched away unwanted copper. Internal circuit layer patterns were protected by a coating of photoresist during etching, while external layer patterns were protected by a tin overplate.

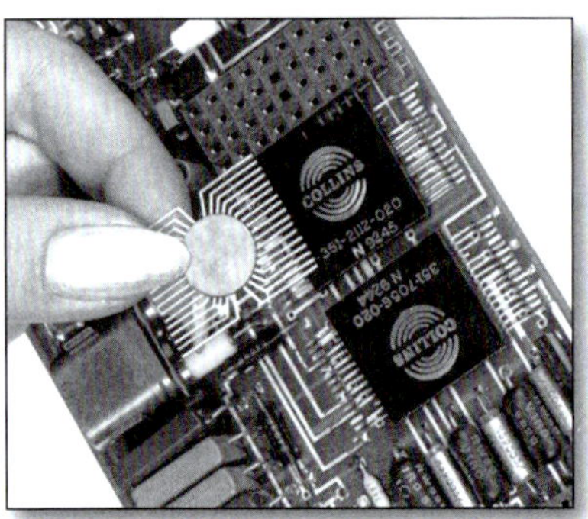

> The MOS/LSI array held between the fingertips replaced the hybrid thin film (shown under array) on this transmit board of a Collins data set.

The announcement of entry into the commercial computer industry followed reports of record earnings and sales for the 1968 fiscal year. Collins noted that the sale of computer services would have minimal effect on fiscal 1969 operating results, because it usually took more than a year to bid for, produce and deliver the equipment.

"Collins is in fact so advanced in computer technology," said the April 24, 1969, issue of the *New York Times*, "that it may be standing on the threshold of a new era.

"Collins Radio's radical approach to computer technology contrasts dramatically to the normal approach. The normal approach, dictated by the inability of present computers to handle information from a variety of points of views, is to look at a single problem at a time — as though boxed apart from everything else. Thus, a single computer user might have to rent time on one computer suited to handling engineering problems, a second one capable of handling automated manufacturing machines, a third to do the billing, a fourth to handle payroll, etc.

> The effort to sell the C-System included an elaborate exhibit booth complete with operating computer systems for demonstrations to convention visitors.

"But Collins has designed a single system to eliminate much duplication and results in one memory bank into which segments of a single computer can 'dip and fuel' at will."

Collins hoped that in addition to developing a market for computer services, the use of these services would lead directly to computer system installations for customers as they developed their own integrated management information, design and production control systems.

To fully develop the market potential, it was necessary for many current computer users to completely "scrap" what they were using and invest in the Collins equipment. Many in the industry doubted that this would happen, but Arthur Collins' previous record of successes in anticipating the needs of the electronics marketplace convinced him, and others, he was right.

But in this instance, time ran out before the market and the product got together. ▪

Off; Denied

...las, Tex., Iowa's

60 The Cedar Rapids Gazette: Wed., Sept. 15, 1969

Collins Sales, Earnings Down; Improvement Seen

...d. Collins ...sday an... pected, the company said. A delay in government procurement programs dealt a blow to the results.

Backlog of orders at the end of the year was $304 million, up from $285 million a year earlier, the company said.

Collins noted its participation as a communications contractor for the successful Apollo moon mission in 1969 was ...

Cedar Rapids Gazette: Tuesday, Dec. 15, 1970

...eductions ...ployment ...llins Radio

...sales and earnings for the year ended Aug. 1 were down sharply.

Sales for the year totaled $450,133,000, down from $447,325,... 000 a year earlier. Net income totaled $5,032,000, down from ...

...the problem when Electronic Data tried taking over Collins with an offer that was theoretically worth 30% more than Collins' recent market price? Simply put: Some big Collins stockholders felt that their $166 million might be worth more in real money than EDS' $597 million. They have a point: For the last nine months, Collins' earnings were $8.3 million vs. $2 million for EDS.

What made EDS worth 300 times earnings vs. 12 times for Collins? One reason: 91% of EDS stock is held by Perot and his board; the high price of the stock is based on a small float and on very little trading. Collins' price was established in a real auction market.

H. Ross Perot

Collins Radio Hit... Electronic Data's Take-Over Offer

Collins Board Indicates It's Weighing Other Propos... From a Number of Fir...

Opposition to Plan Deta...

By Norman Pearlstine
Staff Reporter of The Wall Street Journal

DALLAS—Collins Radio Co. formally rejected a take-over bid by Electronic Data Systems Corp. and indicated they are weighing alternative proposals from a number of ... companies.

McDonnell Douglas Corp. and University Computing Co. are among ... Corp. and University Computing Co. among the companies that have expressed interest in Collins.

Arthur A. Collins, chairman, president and founder of the diversified electronic communications company, mentioned the proposals in a letter to stockholders detailing the opposition to a tender offer for Collins shares by Electronic Data, a small Dallas concern engaged in the design, installation and operation of computer systems.

Mr. Collins said the company has been studying a number of transactions "which, if consummated, might have an important effect on the current situation and the future of ..."

A Collins official added that "more ... companies have approached Collins ... a willingness to assist in solving problems raised by Electronic Data's proposed exchange offer." He would neither confirm nor ...

The Cedar Rapids Gazette

EASTERN IOWA'S LEADING DAILY

10 CENTS

CEDAR RAPIDS, IOWA, THURSDAY, MAY 8, 1969

ASSOCIATED PRESS, UPI, NEW YORK TIMES

Story of the Attempt to Take Over Collins Radio Co.

(This article is published in The Gazette by special permission of the Wall Street Journal in which it appeared May 8, 1969.)

Using the simple device of a common stock tender offer, Mr. Perot planned to swap his stock — selling at some 300 times earnings — for the depressed shares of financially troubled Collins.

Withdrew

But last week Mr. Perot wore a long face. He withdrew his tender offer as suddenly as he had made it, explaining that bank loan agreements of Collins contained a default clause that might jeopardize the proposed take-over.

The terse withdrawal statement hid more that it revealed — so it seems to a close observer of the internal maneuvers of both companies during the five-week struggle. The events of those weeks are a textbook study of how an apparently vulnerable company fended off an unwanted suitor.

At the center of this drama was Arthur A. Collins, 60-year-old chairman and president of Collins Radio. He founded the firm in the basement of his parents' Cedar Rapids, Iowa, home 45 years ago.

An intense scientist, he personally has developed many of the company's electronic patents. "His laboratory is his life," observes one company engineer, who adds that Mr. Collins spent long hours on his pet lab projects even during the thick of the battle with Electronic Data.

Waiting Game

But Mr. Collins' activities were deceptive. He played a careful waiting game and, with unyielding pride, refused to even see Mr. Perot after the tender offer was made.

In the struggle between these two strong personalities, big institutional investors finally threw in their chips with Mr. Collins and assured him victory. Mr. Perot's unwillingness to sweeten the tender offer also contributed to the final outcome.

But victory for Collins Radio hardly seemed assured when the take-over bid was made. Erratic earnings and delayed contract cancellations, coupled with what one financial analyst calls "the most incredible ineptitude for telling its story to Wall Street," had soured many investors on Collins Radio.

Some charged that Mr. Collins had buried himself in research and was oblivious to stockholder interests. The pessimism was reflected in a 50 percent slide in the market value of Collins stock in the last 15 months.

Running a Business

To Mr. Perot, the moment was right for Electronic Data to strike. Munching a cold cheeseburger in his office, he said scornfully that "the chairman's place is in the board room looking out for stockholders, not in the laboratory."

To him, Arthur Collins is "one of the great living geniuses of electronic communications" but a less than adequate executive. "His job is to conduct the symphony, but all he does is play the drums, leaving no one in charge."

Mr. Perot clearly sees himself as a man who can conduct a business. He formed Electronic Data in 1962 after working five years as a salesman for International Business Machines ... chairman, presi... stockholder, with 81 percent of the outstanding shares.

When he went public in September 1968, buyers snapped up 650,000 shares at $16.50 a share — or 118 times fiscal 1968 earnings. The stock soared to $40 by January this year, helped by the fact that Electronic Data earnings last year were 25 times the level three years earlier.

Approached Collins

With his stock at this lofty level, Mr. Perot decided to approach Arthur Collins in January with merger plans. Mr. Perot noted that the market value of his company was close to half a billion dollars — far more than the $150 million value of Collins stock, then selling at $50 a share. Collins' shares were being traded at less than 12 ... from merger when it became clear Mr. Perot would assume control of any combined company.

On Saturday, March 22, Electronic Data informed Collins it was planning a tender offer to Collins stockholders. Two days later, it made public the offer to exchange Electronic Data common stock with a market value of $65 for each Collins share up to a total of 82 percent of Collins stock — with a maximum of 1.5 Electronic Data shares to be swapped for each Collins share.

Agitated

Mr. Collins was livid. For the next five weeks his cigaret consumption was to climb to more than three packs a day from the usual two. But his top aides were even more agitated.

...one or face being taken over," David H. Foster, a vice-president and director, told a reporter at Collins headquarters. W. W. Roodhouse, the company's executive vice-president, who had come into the ... only nod assent ... in silence.

But Mr. Coll... hasty action. "... out a lot of roug... the last 35 ... somehow we'll ... out, too," he sa... reaction was tha... was impossible.

But after thre... steady argumen... lawyers and othe... him the threat wa...

Perot Ju...

Meanwhile, Mr... his associates we... the first week ... tender announce...

24 The Cedar Rapids Gazette: Tues., ...

Collins Sales, Earning... Down; Improvement...

Collins Radio Co. Tuesday announced earnings for the three months ended Nov. 1, 1969, of $2.1 million, or 69 cents per share based on 2,967,427 shares outstanding.

Earnings for the first quarter of the prior fiscal year were $2.3 million, or 81 cents per share based on 2,832,711 average shares. Sales in the current fiscal quarter were $95 million, compared with $112 million a year ago.

E. A. Williams, vice-president, control and finance, speaking at Collins' stockholders' meeting in Cedar Rapids Tuesday, said the lower sales and earnings in the first quarter of fiscal 1969 were primarily attributable to a ...

...reduction in the booking level of contracts during several months.

He noted that the results represent ... low for the fiscal ... followed by ... proving quarters.

Backlog at No... $318 million ... million ...

Des Moines Register Page 13
Thurs., March 27, 1969

OFFICIAL BID FOR COLLINS

Leased Wire From Dow Jones

DALLAS, TEX. — Electronic Data Systems Corp. of Dallas, Wednesday filed with the Securities and Exchange Commission its proposed tender offer to shareholders of Collins Radio Co. of Dallas and Cedar Rapids, Ia.

The company had said Monday that it intended to make a bid for Collins' stock.

Electronic Data, which noted that it already holds 75,000 Collins Radio common shares, said it seeks at least 1,438,388 more to give it at least 51 percent of Collins' 2,967,427 outstanding shares.

...terms of the Electronic Data offer, the company will exchange a maxi...

...oines Register April 1, 1969

THE BUSINESS TIDE

Collins Board Rejects Bid, Hints Other Offers

Leased Wire to The Register

Directors of Collins Radio Co. of Cedar Rapids, Ia., and Dallas, Tex., rejected a takeover bid by Electronic Data Systems Corp., Dallas, and indicated they are weighing alternative proposals from several other companies.

In a letter to stockholders Monday, Arthur A. Collins, chairman, president and founder of the diversified electronic communications company, said the Electronic Data proposal was "hostile to the interests of the company and its shareholders," and that the Collins board "has unanimously decided to oppose vigorously," the offer.

Willing to Assist

The chairman noted that management has been evaluating a number of transactions "which, if consummated, might have an important effect on the current situation and the future of Collins."

A Collins official added that "more than 10 companies have approached Collins indicating a willingness to assist in solving problems raised by Electronic Data's proposed exchange offer." The official declined to elaborate.

Electronic Data, a small concern engaged in design, installation and operation of computer systems, announced last week that it seeks at least 1,438,388 Collins common shares, which with the 75,000 shares it already holds, would give it at least 51 per cent of ...

Honeywell Inc. and Collins Talk Merger

Cedar Rapids firm— ...times those of Collins, ... describes itself as an ...

Merger discussions have been started by officials of ... Radio Co. and Honeywell ... Minneapolis, officials of ... companies announced T...

In a brief statement, ... Chairman James H. Bi... Honeywell and Arthur ... lins, chairman and pres... Collins Radio, the co... said terms of the ... transaction have not y... arrived at.

They said furthe... nouncements will be m... negotiations continue.

Automation Compa...

Tuesday's annou... came on the heels of a take-over of Collins b... tronic Data Systems of ... much smaller compa... EDS stock tender off... nounced two weeks ago, resisted by Collins mana...

Honeywell, which ha... of more than two and ...

WALL STREET JO...

© 1969 Dow Jones & Company, Inc. All Rights Reserved.

MONDAY, APRIL 21, 1969

Collins Radio, Honeywell Talks Collapse, Leaving Collins to Fend Smaller Firm's Bi...

...Co. and Honeywell ... had previously an... ...bid of about $70 a share was strong enough to ... Electronic Data's proposed offer, ... whether Honeywell was willing to increase the ...

...ire ...eal ...with Col-

...Tuesday, Nov. 17, 1970 3

...rst Quarter ...ported by Collins

...year, both in sales and earnings.

"While this particularly exacting period in the nation's economy is a difficult time to make projections, our best judgment presently is that the year's total sales will be slightly below that of 1970 and earnings will see a fair improvement. In making this projection, we have anticipated a low sales volume in the first quarter and its effect ...

Des Moines Register Page 16
Wed., May 28, 1969

COLLINS NET DOWN 29%

Leased Wire From Dow Jones

DALLAS, TEX. — Collins Radio Co., Tuesday coupled announcements of a 29.3 per cent drop in nine-months earnings and a forecast of a "20 to 30 per cent" increase in earnings for the fiscal year ended July 31.

Des Moines Register 3-5
Wed., Nov. 26, 1969

THE BUSINESS TIDE

Collins Radio's Earnings, Sales Off for Quarter

Leased Wire to The Register

Collins Radio Co., large Cedar Rapids, Ia., and Dallas, Tex., manufacturer of electronic equipment suffered a 10.5 per cent decline in earnings for the quarter ended Nov. 1, the first period of fiscal 1969, as sales slipped ...

80-cent dividend rate is 1.2 per cent.

Banks Wer...

(Continued from Page 1)

saw no major bars to total success. And they talked of combining their own knowledge of computers with Collins' broad communications and electronics experience to create a rival to IBM by 1980.

Their optimism was fueled by the quick jump in Collins' stock to more than $60 a share following announcement of the tender offer. More than 272,000 Collins shares were traded in the five days following the announcement, suggesting that institutional investors were buying large quantities of Collins stock because they thought the Electronic Data offer would succeed.

Two weeks after the announcement, Mr. Perot knew that 30 institutional investors held more than one million Collins shares, or one-third of Collins, and that they would hold the key to victory or defeat.

Two days after the offer was made Mr. Perot was predicting privately that he would have commitments from major holders of more than 500,000 shares by the week's end.

There was always the chance that Collins would agree to merge with someone else to thwart Electronic Data. But at ... two nights after the tender offer was announced Mr. Perot discussed this threat ...

ARTHUR ...

...reservations to ... Collins, sources ... apparently ... Justice depart... before Collins ...

Mr. Perot ... spoke with ... John Mitche... Data. But ... friend of the ... worked with ... Nixon campaign ... President Nixon's ... firm for personal ...

In the first week ... nouncement of the ... Mr. Perot's only ... inability to get Arthur ...

...ries to gulp a big one

...ic Data Systems proposes ...Co. Collins opposes the deal, ...ell and other companies

Battle for Survival

A few years ago, Arthur Collins of Collins Radio bet heavily on a new computer. Right now, he appears to be losing that bet.

The Cedar Rapids Gazette

Troubled times

‹ A sampling of newspaper articles from 1969 and 1970 points out the uneasy situation the company faced.

› Collins stock fell from an all-time high of over $100 per share to $50 during the troubling times at the end of the 1960s. Pictured below are Collins Radio Company Stock Certificates from 1959, 1967 and 1968.

The problems which plagued Collins Radio at the end of the 1960s were not confined to one company. The latter half of the decade was very rough on many big defense contractors. Douglas Aircraft almost went under before merging with McDonnell Aircraft. Rolls-Royce Ltd., the maker of engines for Lockheed's new L-1011, declared bankruptcy. That situation strained the position of Lockheed, already reeling from $484 million in losses on four government contracts. Lockheed was in debt to 24 banks which refused to make any more loans to the aerospace giant unless the U.S. Government guaranteed them.

Government defense procurement dropped off as part of a deescalation plan for the Vietnam war. At the same time, space program expenditures were extended or postponed, and some were cancelled entirely.

Major manufacturers of commercial aviation products reduced employment extensively to allow themselves to weather the cutbacks.

Major markets for Collins' products declined, and the company was caught with commitments for major expenditures for plant expansion and product development in anticipation of future needs. Unable to cancel these commitments without incurring extensive cancellation charges, the company had to bear the damaging effects.

Collins had invested millions of dollars in research and development for the C-System, but sales lagged far behind anticipated levels. The practice of amortizing research and development costs the same as capital costs allowed earnings statements to appear in better shape than they actually were. Investors recognized the company's position, and Collins stock fell from its all-time high of over $100 per share to $50, opening the door for one of the most unusual takeover attempts in modern business history.

› In 1971, the State of Iowa selected a Collins C-System for its statewide data network called TRACIS — for Traffic Records and Criminal Justice Information System. After touring Collins facilities, Governor Robert Ray held a news conference to publicize the State of Iowa-Collins Radio venture. Pictured from left: Marvin Selden, state comptroller; Governor Ray; William Roodhouse, Collins executive vice president; and Howard Walrath, Collins general services vice president.

Little fish vs. a big fish

On March 24, 1969, H. Ross Perot stunned the business world by announcing that his small Dallas-based computer concern, Electronic Data Systems Corporation ($8 million annual sales), intended to take over Collins Radio Company ($440 million annual sales).

Perot boasted that he was about to pull off the classic takeover. Using a common stock tender offer, he planned to swap his stock, selling at 300 times earnings, for the depressed shares of Collins. Electronic Data announced that it sought at least 1.4 million Collins common shares to add to the 75,000 it already held, to give it 51 percent of the Collins outstanding shares.

Electronic Data would exchange common stock having a market value of $65 a share for each Collins common share tendered with the maximum of 1.5 Electronic Data shares being swapped for each Collins share.

What followed was just the beginning of a struggle between two strong personalities — Arthur Collins and H. Ross Perot.

Two months earlier, Perot had approached Collins with merger plans. Perot noted that the value of his company was close to half a billion dollars — compared with the $150 million value of Collins stock. Collins was impressed with Perot, but he backed away from the merger when it became apparent that Perot would assume control of any combined company.

Since his "friendly" method of merger did not work, Perot made the tender offer for the Collins stock. Top aides of Collins were concerned that the offer might force the company to combine with some other company to avoid an unfriendly takeover. But Mr. Collins opposed any hasty action. His initial reaction was that a takeover was impossible.

"We've ridden out a lot of rough storms in the last 35 years," Collins said, "and somehow we'll ride this one out, too."

> A. H. Gordon with Arthur Collins. In addition to his chairmanship of Kidder, Peabody & Co., Gordon also served on the Collins Radio Co. board of directors.

But after three weeks of steady argument, company lawyers and others convinced him the threat was real.

Optimism characterized Electronic Data's headquarters. Collins Radio stock jumped to more than $60 a share following the announcement of the tender offer. High levels of trading indicated institutional investors were purchasing large amounts of Collins shares in anticipation of a successful offer. Two weeks after the announcement, Perot knew that ten institutional investors controlled more than one million shares of Collins common stock. Those investors held the key to victory or defeat.

Collins, aided by A. H. Gordon, chairman of Kidder, Peabody and Co., and a long-time friend of Arthur Collins, searched for potential "white knight" merger possibilities to thwart Electronic Data's takeover attempt. Exploratory talks were started with Harris-Intertype Corp., Burroughs Corp., Control Data, University Computing Co., and McDonnell Douglas.

Sam Wyly, president of University Computing, agreed to buy 300,000 shares of Collins common, but had to give it up when his own company's stock fell. Control Data ran into formidable Justice Department reservations to any merger with Collins.

These developments seemed to further Electronic Data's probability of success, but in the week following the tender announcement, Perot was not able to get Arthur Collins to meet with him to discuss the deal. Perot knew his chance of success would improve if he could get Collins and other board members to publicly accept the tender offer. The crucial question seemed to be what the big institutional shareholders would do if Mr. Collins walked out of any discussions.

During the second week following the offer, Collins publicly rejected the Electronic Data tender offer. In a special letter to stockholders and in a large ad in the April 3, 1969, *Wall Street Journal*, Arthur Collins stated the Electronic Data proposal was "hostile to

To the Stockholders of Collins Radio Company

Your Company's Board of Directors has unanimously decided to oppose vigorously the exchange offer proposed by Electronic Data Systems Corporation ("Electronic"). None of the Company's management or directors will accept the Electronic proposal.

We believe that the proposal is hostile to the interests of the Company and its stockholders. This belief is derived from the following considerations:

(1) Based on the latest fiscal earnings and the over-the-counter bid quotations supplied by Electronic in its registration statement, the price earnings ratio of its stock is in excess of 300 times. Of the outstanding Electronic shares, 91% are owned by nine officers and directors and additional shares are owned by other employees.

(2) You are being asked to make an exchange for stock which was first issued to the public about six months ago. The Electronic stock is not traded on the New York or any other stock exchange and there is no public information regarding the volume of trading.

(3) The latest reported fiscal year's sales of Electronic were only $7,666,000 as compared to Collins' $447,026,000 and its latest reported earnings per share for such period were 14¢ as compared to $4.44 for Collins.

(4) As set forth in the Electronic registration statement, Collins' shareholders accepting the offer would suffer very large dilution in earnings per share. For example, the earnings represented by each Collins share exchanged for 1.5 Electronic shares would be reduced from $4.44 to 25¢ if 20% of Collins' shareholders should exchange, and 93¢ even if 51% should exchange.

(5) Electronic has never declared any dividends and states that it does not expect to grant cash dividends in the near future. In the past twelve months, Collins' shareholders have received cash dividends of 80¢ per share.

(6) Acceptance of the Electronic exchange offer will be a taxable transaction to Collins' stockholders.

(7) Tenders to Electronic will be irrevocable and, therefore, Collins' stockholders lose control of their stock once the shares are tendered. Electronic, on the other hand, is not bound to accept any shares unless 51% or more are tendered.

(8) The Electronic proposal raises significant questions under the Federal AntiTrust laws, could have a serious adverse effect on the relations between the Company and its major customers (including the United States Government) and could jeopardize the personnel relationships within the Company which are vital to its continued technical achievements.

Your Board of Directors has great confidence in the future of Collins Radio Company, particularly in view of our recently announced intention of fully realizing the potential of our communication-oriented data systems as a major sector of business. The Company's research and development expenditures during the last twelve months for this and other future programs far exceed the total revenues of Electronic for its most recently reported fiscal year.

Your management has been evaluating a number of other transactions which, if consummated, might have an important effect on the current situation and the future of Collins Radio Company. We will keep you apprised of all pertinent developments in this regard.

Very truly yours,

Arthur A. Collins, President
By Order of the Board of Directors

March 28, 1969

‹ This letter was mailed to Collins stockholders and appeared in the April 3, 1969, *Wall Street Journal.*

the interests of the company and its shareholders," and that the Collins board of directors "has unanimously decided to oppose vigorously" the offer. Collins added: "Based on the latest fiscal year earnings and the over-the-counter bid quotations supplied by Electronic Data in its registration statement, the price-earnings ratio of its stock is in excess of 300 times."

Meanwhile, Collins looked around for other potential merger partners to save itself from Electronic Data. On April 8, 1969, two Collins vice presidents flew to Minneapolis to talk with Honeywell, Inc. But both sides saw Justice Department objections, and they couldn't agree on Mr. Collins' role in any merger. The Collins board ended merger talks with Honeywell April 19.

When an Electronic Data vice president heard the Collins-Honeywell talks had been called off, he was reported to have said, "The ball game is over," reasoning that time was running out for Collins.

But Mr. Collins held out. Four weeks after the disclosure of the offer, Chase Manhattan Bank, holder of 455,000 Collins shares, voiced its opposition. Chase and another New York bank, Morgan Guaranty Trust Co., were leery of swapping their shares for the high-multiple Electronic Data stock. They were, according to reports, waiting for Electronic Data to raise its offer to about $100 worth of stock for each Collins share. That presumably would have driven Collins stock up to approximately $85 a share, or high enough for the banks to unload their holdings at a profit.

Perot refused to sweeten the original offer. Chase Manhattan and other large holders consolidated their opposition and sided with Collins. Electronic Data withdrew the tender offer before it was ever approved by the Securities and Exchange Commission.

An editorial in the May 11, 1969, *Cedar Rapids Gazette* reflected on the sequence of events:

"If *Wall Street Journal* reporter Norman Pearlstine's account of the recent abortive attempt to take over the Collins Radio Company is reasonably accurate — and we have no reason to doubt it — the people of Cedar Rapids and its surrounding area can add one more to many other reasons for being grateful to Arthur Collins. Evidently his cool determination and rugged spirit spared thousands of people in this community a great deal of prolonged anxiety, and possibly worse.

"We don't need to be financial sophisticates to sense how much Collins Radio means to all of us whose lives are rooted in this community. So it was not pleasant to watch from the sidelines an uncommonly audacious attempt to restructure an industry so important to the Cedar Rapids economy against the resistance of the man who built it."

Eighteen months later, Collins announced that it had initiated exploratory talks with TRW, Inc., about possible affiliation, but within two months, the talks were discontinued.

Rockwell enters

The critical cash flow condition at Collins continued to plague the company's operations in 1970. The company was in violation of cash conditions of several lenders' agreements, and sales continued to drop with the worldwide aerospace recession.

› Willard F. Rockwell, Jr. (left), chairman of the board of North American Rockwell, and Arthur A. Collins signed the agreement September 2, 1971, for the $35 million investment in Collins Radio.

The Anamosa plant was closed in January, 1971, and the 240 employees were transferred to the Main Plant in Cedar Rapids.

The day following discontinuation of talks with TRW, it was announced that Collins and North American Rockwell Corporation had begun discussions on a possible significant investment in Collins. Two weeks later, on June 1, 1971, the managements of the two corporations had agreed on the terms of North American Rockwell's investment in Collins.

Subject to approval of respective boards of directors and Collins shareholders, North American Rockwell would purchase $35 million of a new issue of Collins convertible preferred stock, and would have warrants to purchase an additional $30 million of Collins common stock. The initial conversion and warrant exercise price was set at $18.50 per share, and the convertible preferred stock would give North American Rockwell the right to elect a majority of Collins' 13-member board of directors.

On June 10, the boards of directors for the two firms approved terms of the proposed investment. In a letter to Collins stockholders, Arthur Collins urged approval of the sale.

"The company's ability to continue its business clearly is dependent upon the receipt of substantial new financing. The proposed investment by North American Rockwell provides the needed additional capital and also does more . . . the participation by North American Rockwell's managerial talent on Collins' board of directors, while not necessarily assuring any improvement in Collins' present financial condition, will afford invaluable counsel in many aspects of our operations."

Pointing out that directors of both companies favored the proposal, Collins said its investment bankers also recommended approval of the agreement.

"Collins believes that if the proposed transaction with North American Rockwell is consummated, the proceeds . . . together with the restructuring of the bank debt . . . would enable Collins to meet its current financing requirements."

› After a meeting of the new board of directors on November 23, 1971, it was announced that Robert Wilson had been named new president and chief executive officer of Collins. At the stockholder's meeting, the election of directors representing Collins stockholders was postponed for a month. Seated at the head table were, from left, Donald Beall, Collins senior vice president for Finance and Administration; Willard Rockwell, Jr., newly-elected chairman of the Collins board; Robert Frost, Collins vice president and company secretary; and James Lenehan, Collins vice president of Telecommunications and Data Marketing.

Collins stockholders approved the proposal on August 31, 1971. The agreement was consummated September 2, 1971, with the investment of $35 million in exchange for the agreed upon Collins shares and warrants.

North American Rockwell's original plan was to leave Collins management intact. But when the fiscal year 1971 loss of $46 million was followed by an $8 million first-quarter loss in fiscal year 1972, Willard F. Rockwell, Jr., chairman of North American Rockwell, announced that Robert C. Wilson had been named president and chief executive officer of Collins, with Arthur Collins to continue as a member of the board. Other board members elected were R. L. Cattoi, senior vice president; W. C. Hubbard, vice president and controller; J. D. Nyquist, senior vice president; H. M. Passman, vice president; and W. W. Roodhouse, executive vice president.

When Collins was replaced as president and chairman of the board of the company he established, Roodhouse was traveling abroad. When he returned and discovered Collins had been replaced, Roodhouse submitted his resignation. Within a week, he returned to the firm convinced that the change in management was necessary for the company's survival.

On January 14, 1972, Collins announced his resignation from the board of directors. He announced the formation of a new corporation, Arthur A. Collins, Inc., to carry out system engineering studies in communications and computer fields.

Wilson, the new Collins president, had been with General Electric some 25 years before being tapped to head the commercial products group of North American Rockwell in 1969. At Rockwell, Wilson played a major role in recommending the investment in Collins. His recommendation was based largely on the fact that the two markets served by Collins — aviation and communications — were two of the very fast growing areas of the electronics industry.

"If you look at what's happening in communications, and what will happen over the new couple of decades, it is going to be almost an explosion on the total world base," Wilson said. He also believed that Collins Radio was rich in intangible assets such as its outstanding reputation with customers, market position, leadership in technology, and outstanding personnel.

"When things continued to look bad," Wilson said, "Rockwell management suggested that perhaps I ought to follow my judgment and come down and try to run it. So I did."

His first priority was to assure Collins personnel about the future of the business and the company. He also had to do a great deal of fence-mending with bankers and creditors.

Arthur Collins preferred the solitude of his laboratory to meeting with investors, creditors and customers. While he toiled in the lab, he sent subordinates to deal with the bankers.

Wilson, on the other hand, was described as the "hard-driving, outgoing, managerial type." He frequently held meetings with Collins officials in airports while he waited to change planes, talked freely with bankers, had breakfasts with local reporters, and made numerous speeches.

> Robert C. Wilson

Wilson's plan

Upon arriving at Collins, Wilson sized up the situation and decided that all the things which led North American Rockwell to invest in Collins were there. However, he recognized the presence of several problems. Employee attitudes were poor because the company's future appeared questionable. The company showed a net loss of more than $40 million for the 1971 fiscal year, and things looked just as bad for 1972. Cash drain was projected to increase while orders continued to lag. The company needed strict fiscal controls, and many programs were in trouble. Wilson also recognized that the company needed better strategic planning. He felt that though these problems were deep-seated, they could be solved.

Wilson began his recovery plan with major cost-cutting policies. Capital expenditures were completely eliminated, and research and development funds were cut back to a level more in line with sales. Long-distance phone calls were reduced to a minimum, lights were turned out, use of copy machines was limited, duplication of research efforts was terminated, and 600,000 square feet of leased space was vacated. Major cuts were made in personnel, with 2,000 engineers and administrators cut from the payroll. The company also improved its cash position through the sale and leaseback of $7.8 million of company-owned equipment. Additional funds were received from a similar transaction of some of the company's production equipment.

The company changed its accounting practice from deferring development costs over product deliveries to expensing them as they occurred. A $36 million inventory write-down was made to get the value of such assets more in line with sales expectations. These two moves alone accounted for a major portion of the company's 1972 fiscal year loss of nearly $64 million. Wilson and his assistants conferred with Collins' lenders in 1971 and negotiated amended loan requirements to allow the company to get back in the black.

Coupled with the tight fiscal control was Wilson's major restructuring of the company's management in December, 1971. Nine organizational levels were reduced to five in some areas. The company was decentralized into four major operating groups: the Avionics and Telecommunications Group in Cedar Rapids; Telecommunications and Switching Systems in Dallas; the Special Telecommunication Group in Newport Beach; and International Operations headquartered in Dallas. Each group in turn was divided into operating divisions (totaling 15) which Wilson optimistically called profit centers. Each division concentrated on a single product area from marketing through engineering to production, to increase awareness of the bottom line.

Other major actions taken included reevaluation on pricing policies and the establishment of new prices on many products, all with the approval of the Federal Wage and Price Board.

As part of the new organizational structure, marketing was emphasized as a critical factor for future company success.

The initial plan was to keep the company's focus on such traditional products as avionics and telecommunications systems, to expand its development in the growing

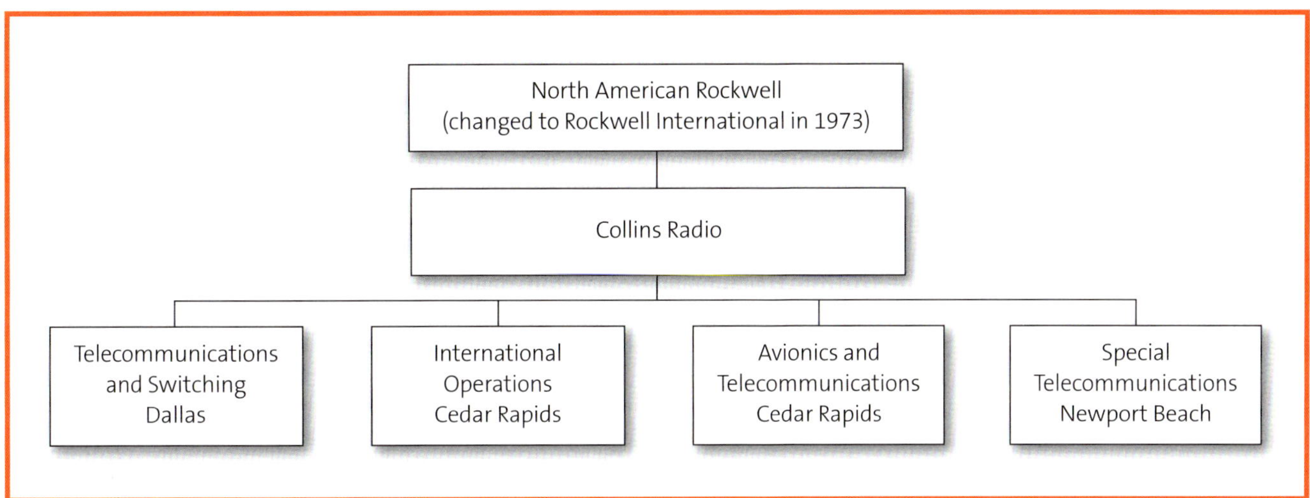

> Under Wilson's plan, Collins was decentralized into four operating groups in 1971. Note that this chart does not reflect present groupings of operations.

microwave business, and to shift emphasis away from computer and semiconductor activities where Collins was having a tough time with the competition.

The company's five-year plan, as expressed by Wilson, was to develop Collins into a solidly profitable and diversified growth company.

Part of the plan was to continue the development of international markets. In 1972, approximately one-third of total volume was achieved outside the United States. By 1977, the company expected that portion to be one-half.

In a presentation to security analysts in April, 1972, Wilson said he was confident the outlook for the future was good.

The turnaround

In March of 1972, the first 100 employees were called back to work. A March 17, 1972 editorial in the *Cedar Rapids Gazette* praised the development.

"There is every reason to conclude that negatives of paralytic gloom accumulating from the last two years are warranted no more on anybody's part."

The writer of the editorial was correct. Orders and sales for the first half of 1973 showed marked improvements of 66 percent and 46 percent, respectively, over the same period in 1972.

After reporting a $1.1 million profit for the first quarter of 1973, Willard F. Rockwell, Jr., chairman of the board of Collins Radio, extolled the efforts of Collins employees: "In my 37 years in business I have never seen a turnaround as dramatic as this!"

From an operating loss of $17 million in 1972, Collins made a profit of $3 million in 1973. The company reduced its assets employed by 15 percent while sales increased by nearly 50 percent.

The profitable 1973 fiscal year led many investors to believe the merger of Collins Radio and North American Rockwell (renamed Rockwell International in February, 1973), was just a matter of time.

On August 20, 1973, Rockwell made a tender offer to purchase any and all shares of Collins common stock at $25 per share. Rockwell acquired 75 percent of Collins' outstanding common stock for approximately $57 million, and the merger of Collins into Rockwell was approved at a special meeting of Rockwell stockholders on November 2, 1973. Additional costs to acquire the balance of outstanding common stock and to consummate the merger came to approximately $19 million. ▪

> The merger of Collins Radio Company into Rockwell International was approved by Collins shareholders at a special shareholders meeting held in Cedar Rapids on November 2, 1973. Robert Wilson called the merger "The end of one phase in the development of Collins and the beginning of another which promises to be even more successful and dynamic than the first."

FLY COLLINS

Jewel in the crown

"Progress may be too mild a word to describe the Collins about-face," a reporter wrote in the June 17, 1973, *Cedar Rapids Gazette.*

"Certainly the improvement in the economy has helped us a lot," said Clare Rice, senior vice president and general manager of the Collins Avionics Division. "We're in the aviation industry, and both the air transport sector and business aviation sector suffered in the recession for two or three years. But many other things helped too. We are now oriented to our markets, and our business teams are devoted to specific areas much more so than before. We are decentralized and I'm sure the focus we now have on various business areas is a significant factor in our financial performance."

‹ A Gulfstream I company airplane sits beside the Collins hangar at the Cedar Rapids Airport. This airplane was sold in 1983. (*See related article in this chapter.*)

› Communication Switching Systems (C-Systems) played a key role in helping Collins with profitable growth during the mid-1970s. Pictured below is a Collins C-900, a medium-size computer-controlled message and data switching system.

The leaner Collins team continued its profitable growth in the mid-1970s and contributed significantly to the commercial and international business outlook for Rockwell International. Orders of about $500 million for 1974 represented an increase of 60 percent over the 1972 level; and more than 16 percent over 1973. Nearly 60 percent of sales were to commercial and international customers.

One aspect of the turnaround was a new marketing strategy for the C-System.

When financial analysts asked Wilson what he intended to do with the C-System, he replied, "You tell me what C-System means and I'll tell you."

That uncertainty did not last long. Wilson and Donald Beall, executive vice president of the Collins Radio Group, focused the C-System on a specialized but fast-growing segment of the communications market — voice and data switching. The C-System provided a good technical foundation for designing telephone exchanges and data switches for computer networks.

To get its foot in the door, the company turned to the airlines — long-time customers who were experiencing problems in data communications and telephone switching.

"The engineers from Collins really did their homework," said one airline official. "They came to see what we needed." The Collins engineering team worked with Continental Air Lines

> Employees present at the 1974 shipment of the first Galaxy system were, from left: Jim Miller, Gayne Ek, Jim Rise, Dean Thomas, Bob Abbott, Bob Bent, Harvey Lembke, Vivian Cerney, Bob Hirvela, Francis McMann, Clarence Marshall, John Pollpeter, Les Lauther and Bob McArthur.

to design totally new systems to handle both telephone and data communications traffic for the airline's vital phone reservation and ticketing systems. Systems were later installed for Pan American and United.

Telephone companies offered automatic call distribution with their Star system, but because the Collins system offered so many more features, it was named the Galaxy system.

It was the first time such a sophisticated telephone system was used to solve the problem of evenly distributing large volumes of incoming calls to banks of telephones. It could also report on the number of calls coming in, how many callers were waiting, and how long they were waiting, among other features.

Once other companies saw the system working, more orders came in. By 1976, more than 25 Galaxy systems were on order, mostly for domestic airlines, hotel reservations systems, car rental companies, and credit authorization agencies.

The profit picture for other Collins markets also improved.

(*In March, 1974, Robert Wilson resigned as president to become chairman of the board, president and chief executive officer of Memorex Corporation. Succeeding Wilson as president of the Collins Radio Group was Donald R. Beall, who had been executive vice president since December, 1971.*)

> Donald R. Beall, 1982 photo. Beall became president and chief operating officer of Rockwell International in 1979.

Avionics resurgence

A key market which returned to good health in the 1970s was aviation. Highlighting the growth was a 30 percent increase in avionics sales to general aviation and international customers between the 1973 and 1974 fiscal years.

In the early 1970s, a new integrated circuit process known as MOS (metal oxide semiconductor) became practical for avionics. The equivalent of hundreds (and even thousands) of transistors were processed on a single chip. MOS structures, smaller and requiring fewer process steps than the bipolar integrated circuits of the early 1960s, became the workhorses of the new avionics technology.

At the heart of this new era was the use of digital processing to fly the aircraft, manage fuel consumption, monitor engine performance, control communications and perform such navigational functions as storage and retrieval of waypoints, VOR-DME tuning and course and speed computations.

One application of the new technology was area navigation, or RNAV. A concept long recognized for its potential, RNAV provided direct point-to-point navigation off the established airways.

The new generation wide-body jet transports were the first to used advanced area navigation. (*See Chapter 8.*)

By installing these new digital systems aboard aircraft which were otherwise equipped with analog instruments, the aviation industry took the first step toward the all-digital age of the 1980s.

Collins also made technological strides in the business aviation field during the 1970s. The first of a new line of avionics was introduced in 1970 for business jets

> Automatic call distribution systems were used by every major American airline and several foreign air carriers for computerized routing of telephone calls. Ozark Airlines installed a system at its regional reservation center in Peoria, Illinois.

Ozark photo.

and twin-engine airplanes. Low Profile avionics featured remote-mounted boxes which were half the height of standard boxes and could be located in places previously too cramped for avionics. The height was the smallest of any remote box in the industry and allowed the installer, if he wished, to stack two boxes in one vertical space of the customary radio rack.

In the fall of 1973, Collins introduced its RNAV and control system for business aviation. The NCS-31 was a digital area navigation computer that also included an optional touch tune central control for all navigation and communication radios.

A Collins-owned Beech Duke criss-crossed the nation in 1974 and 1975 to demonstrate to business aviation customers the quality and versatility of the new avionics. Later tours of Europe and South America showed the world the reliability of the Low Profile Line.

The Low Profile name was changed to Pro Line in 1975, and by the early 1980s, more than 100,000 units had been sold, making Pro Line one of the world's most successful product lines in business aviation.

In 1975, a new line of panel-mounted avionics for the single-engine and light twin-engine market was introduced. Micro Line products were based on the latest electronic display and microelectric circuits.

› The first 40-year anniversary pin presented to an employee went to Katherine Horsfall in October 1973. Miss Horsfall joined Collins Radio Co. October 3, 1933 and served continuously in secretarial positions since that date. Presenting the pin were Clare Rice (left), who was named president of the Collins Avionics Group in 1977, and James Churchill, who succeeded Rice in 1981.

› Collins Pro Line was an avionics line designed for twin-engine airplanes and business jets — aircraft which most often would be flown by professional pilots. 1978 photo.

> Reams of paper made up the TACAN proposal for the U.S. Air Force in 1974. Ed Baermann, Ann Bunting, Shirley King and Thad Kuenz are shown sorting the 28,000 sheets reproduced by the company's Reproduction Center.

TACAN contract

One of the major Collins products during the 1970s and 1980s began with a significant contract award in August, 1973.

The Collins organization of Cedar Rapids was contracted to develop a new generation of TACAN equipment to meet new 252-channel requirements of the military. At that time, TACAN units were large and relatively unreliable compared to other equipment, and some of the technological advances in other avionics products had not been applied to TACAN.

Consequently, at the beginning of the 1970s, the U.S. Air Force established a list of features for a new system. The new TACAN had to be a smaller, lighter unit using advanced design techniques. It had to double the number of radio channels, add ranging capability, have high reliability and cost less than $10,000.

In April 1973, Collins was one of two companies each awarded a $1.5 million contract to develop and produce seven TACAN units for test and evaluation. The Collins team knew that if it won the competition, the Collins TACAN would likely be designated as standard equipment for the Air Force, and orders could total 10,000 units.

In July, 1975, the Air Force announced its decision. The team was awarded an initial production contract totaling $14 million for 1,000 systems, with options for an additional 7,000.

Iowa Senator John Culver praised the Collins TACAN program on the floor of the U.S. Senate in 1978.

"The results to date have exceeded everyone's expectations," the senator told his colleagues. "In qualification tests, the Collins TACAN achieved a reliability of about 1,000 hours mean time between failures, which was double the original specifications. In field use, according to the most recently available data, the mean time between failures is more than 1,800 hours — or more than triple the requirement."

> AN/ARN-118(V) TACAN

With the original Air Force TACAN award, the division pioneered the Reliability Improvement Warranty concept for military avionics. Under RIW, guaranteed field performance and a comprehensive maintenance program were provided to support equipment delivered to end customers.

The Collins Government Avionics Division achieved a milestone in 1979 when it shipped the 10,000th AN/ARN-118(V) TACAN, bringing the value of total orders to more than $125 million. The ARN-118 is one of the more successful products ever built by the division.

Reorganization and expansion

By the summer of 1974, employment in Cedar Rapids was climbing at a rate of 100 persons each week as the backlog of orders continued to increase.

Business Week magazine quoted Willard F. Rockwell, chairman of the board of Rockwell International, as saying the former Collins Radio Co. was "the jewel in the crown" of the Rockwell organization.

In conjunction with its decentralization efforts, Rockwell management organized the Cedar Rapids Collins operations into five distinct divisions in 1977.

The Collins Telecommunications Products Division (CTPD) was chartered to manufacture radio communication products.

The Collins Communications Switching Systems Division handled automatic call distribution systems.

The three avionics divisions — Air Transport, General Aviation, and Government — comprised the Collins Avionics Group. Clare Rice was named president of the Collins Avionics Group.

In June, 1974, the company leased a manufacturing plant in Melbourne, Florida, while construction began on a new, larger building in Florida. Located on that state's east coast near Cape Canaveral and the Kennedy Space Center, Melbourne offered a new source of trained electronic personnel close to a hub of aerospace activity.

In 1977, a new 40,000-square-foot facility was dedicated in Melbourne for the Collins General Aviation Division.

Also in 1977, four expansions in Iowa were announced.

A mechanical assembly plant employing about 40 persons opened in Manchester for the Collins Air Transport Division.

In Anamosa, the Collins Telecommunications Products Division opened a service center in the building used by Collins as a manufacturing plant from 1955 until 1971.

A new 42,000-square-foot facility opened in Mason City for inspecting purchased parts and fabricating plastic parts for Collins avionics products.

In Cedar Rapids, Collins Communications Switching Systems Division began construction of a 110,000-square-foot building for laboratories, manufacturing, warehousing and office space on a 10-acre site near the C Avenue complex.

The following year, Collins Air Transport Division opened its new 7,500-square-foot facility in Decorah, Iowa.

Expansion continued in Cedar Rapids in 1978. Administrative offices of CTPD were moved to leased space in the Life Investors Inc. building, and more than 100 of the division's employees occupied the second and fourth floors of the Iowa-Illinois Gas and Electric Co. building. The company's Graphic Services Department moved into leased space in the former Vigortone Products headquarters on Blairs Ferry Road.

› The new Melbourne, Florida, plant was dedicated in 1977.

> The Advanced Technology and Engineering Center (Building 124) in Cedar Rapids, as viewed from the east.

> The atrium of Building 124 in Cedar Rapids serves as a reception area for the Avionics Group.

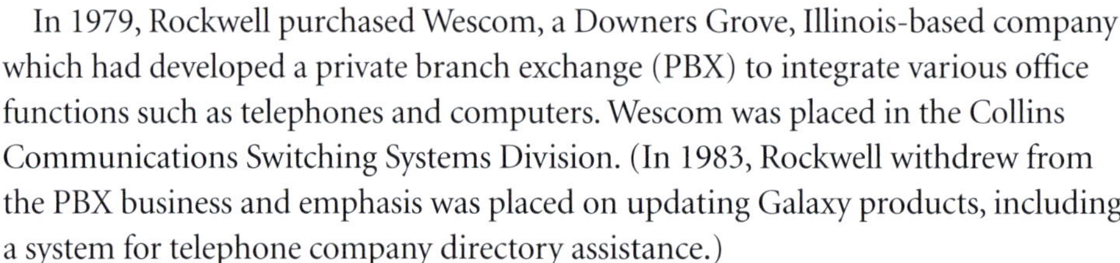

In 1979, Rockwell purchased Wescom, a Downers Grove, Illinois-based company which had developed a private branch exchange (PBX) to integrate various office functions such as telephones and computers. Wescom was placed in the Collins Communications Switching Systems Division. (In 1983, Rockwell withdrew from the PBX business and emphasis was placed on updating Galaxy products, including a system for telephone company directory assistance.)

"Advanced technology and engineering capability is the heart of design and development of new avionics products and systems for the markets we serve," Rice said in announcing a new expansion. "And these markets are growing rapidly. Consequently, we've run out of available laboratory and office space."

In late 1979 and through the spring of 1980, about 800 avionics engineering employees moved into a new 200,000-square-foot engineering and research facility at the C Avenue complex in Cedar Rapids (Building 124).

A new facility for automated warehousing and assembly for CTPD was opened in Salt Lake City, Utah, in 1979.

Another expansion was made in 1980. Two Rockwell divisions with similar customer interests — the Collins Government Avionics Division of Cedar Rapids and the Missile Systems Division of Columbus, Ohio — opened a 75,000-square-foot plant near Atlanta, Georgia, to produce a variety of electronic assemblies.

Major contracts

During the late 1970s, several large contracts were awarded to the Collins divisions in Cedar Rapids, as the telecommunications and aviation industries strengthened.

Among them were the new standard U.S. Air Force transceiver, the AN/ARC-186(V); a $48 million avionics package for the U.S. Coast Guard's Medium Range Surveillance (MRS) aircraft; a Short Range Recovery (SRR) helicopter avionics package for the U.S. Coast Guard; and a new high frequency radio for about half the U.S. Air Force's aircraft (ARC-190).

> Jack Cosgrove (right), vice president and general manager of Collins Telecommunications Products Division, accepts a trophy model of an early ARC-159 radio from Tom Wilford, division manufacturing manager. The event marked the 5,000th such unit delivered to customers. 1977 photo.

Autopilot proves its value

A 34-year-old helicopter pilot involved in a mid-air collision over a major metropolitan area credited the Collins autopilot system installed in his aircraft with saving his life.

"There's no doubt about it, none whatsoever in my mind," says Edward Gabryszewski of Sterling Heights, Michigan. "Without the Collins autopilot, I'd probably be dead right now."

On March 22, 1982, Gabryszewski, a pilot for WJBK-TV (CBS) in Detroit, dropped off a news crew at the television station and took off for nearby Berz-Macomb Airport to secure his helicopter for the night.

After clearing the towers and tall buildings of downtown Detroit, he activated the Collins APS-841H autopilot in the Bell JetRanger III helicopter and started to relax.

Thirty seconds later, by his own estimation, Gabryszewski was knocked unconscious by what he describes as "a loud explosion." The "explosion," he would learn at the end of his ordeal, was a five-pound mallard duck crashing through the JetRanger's windscreen.

In the collision, the mallard "took out two-thirds of the windscreen on the pilot's side and struck me square in the face," Gabryszewski recounted. "The force of the impact broke my nose but the duck, for the instant it was on my face, shielded my eyes from the flying plexiglass that followed it through the hole."

Gabryszewski was unconscious for 90 seconds. Revived by the cold wind blowing through the JetRanger's shattered windscreen, he found himself still cruising at 700 feet and 110 knots over Detroit.

"When I came to, I was still in my harness, slumped over the co-pilot's seat. I tried to sit up. But as I did, I grabbed the cyclic and accidentally disengaged the autopilot. The chopper started to climb immediately and began to do an inverted roll to the right. I was only semi-conscious, but I could feel what was going on, so I started fighting like hell to regain control. Then I realized I could get some help. I punched up the autopilot again and let go of the cyclic. The chopper came back around. I leaned back and cleared my head."

Gabryszewski managed to safely land the helicopter at a nearby airport — Big Beaver — and was joined there minutes later by a concerned police helicopter pilot. It was the latter who discovered the cause of the "explosion," lying atop a videotape recorder in the JetRanger's rear seat.

Now fully recovered from his injuries, Gabryszewski considers himself "very fortunate, very lucky. I still can't thank Collins and my president enough... for my life."

It was Kenneth Bagwell, president of Storer Broadcasting Co. of Miami, which owns WJBK, who authorized installation of the APS-841H at Gabryszewski's request. ▪

› Pilot Ed Gabryszewski poses with five-pound mallard duck that crashed through the windscreen of the helicopter he was flying over Detroit. 1982 photo.

The first active control system, for the Lockheed L-1011-500 jetliner, was delivered in 1979. The system, combined with nine extra feet of wing span, was designed to reduce drag, fuel consumption, and structural wing loads. The application was the first commercial use of active controls technology and remained the only such system in use by 1983.

The commitment to America's space program continued in the 1980s when Collins navigation systems were provided for the Space Shuttle.

> Left: The advanced Collins flight management system (FMS-90) is used to program automatic navigation on board the aircraft, and can control several other avionics systems. 1981 photo.

> Right: Flight deck of the U.S. Coast Guard short-range recovery helicopter featured a four-axis autopilot, dual displays, and dual cockpit management system control and display units.

New navigation equipment

Collins produced one of the earliest automatic navigation systems, the ANS-70, for the large commercial jets in the early 1970s, but its large size and price tag weren't suited to the needs of smaller business aviation aircraft.

In September, 1979, Rockwell International purchased Communications Components Corporation, a Costa Mesa, California-based electronics firm. CCC's product line consisted of simple antennas and highly sophisticated long-range navigation radios. The acquisition was made to expand the Collins general aviation avionics product line.

Collins engineers took the long-range navigation idea a step further by adding complete flight management capability and calling it the FMS-90.

So many cockpit functions could be controlled by the FMS-90 that when the system was introduced at the 1980 National Business Aircraft Association convention, one aviation writer asked, "Can it brew coffee?"

In July, 1979, the Collins Government Avionics Division received a $68 million contract for a totally new type of navigation equipment — the largest developmental contract Collins had received since Project Apollo.

The new Navstar Global Positioning System (GPS) was viewed as the navigation system for the future, intended to ultimately replace many of the navigation methods used by aircraft, ships and land vehicles.

The system, when fully implemented in the late 1980s, would consist of three segments.

The space segment was to be made up of 18 satellites operating in three separate orbits.

The ground control segment would consist of stations to track the satellites, monitor their operation and make corrections to insure accuracy.

The user segment, to include Collins-developed equipment, would receive the satellites' signals and convert them into a precise position readout. The system was to be so accurate that a pilot, the captain of a ship, or even a foot soldier carrying GPS equipment would be able to determine, in three dimensions, his location anywhere on Earth within a few meters.

After nearly a decade of program involvement, the Collins Government Avionics Division decided to demonstrate the potential of the system for civil users. Collins

› Navstar Global Positioning System equipment developed by the Collins Government Avionics Division.

engineers placed GPS equipment in a Rockwell Sabreliner business jet in May, 1983, and made the first transatlantic flight using signals from Navstar satellites for navigation. The Collins crew landed at Le Bourget Airport, site of the Paris Air Show, and then relied on precise satellite positioning data to taxi the aircraft within eight meters of a predetermined parking point.

Collins engineers also worked on a development contract with General Motors to install the sophisticated GPS navigation equipment in automobiles.

The Collins Government Avionics Division was one of two manufacturers vying for the full-scale GPS user equipment production contract, expected in 1984.

Picture tubes in the cockpit

Television picture tubes — cathode ray tubes (CRTs) — have been around since the 1940s, but their use in aviation was limited for many years. The early tubes were heavy and required complex circuitry. Their biggest drawback was a lack of picture brightness.

The 1950s brought brighter CRTs, and single-color displays began to be used in cockpit weather radars. Advances in tube design and integrated circuitry through the 1960s, plus the advent of color displays, finally made CRTs economical and practical for broader applications in aviation.

By the early 1970s, Collins engineers were drawing on their expertise in flight display technology to begin experimenting with CRTs as primary flight instruments — taking the place of electromechanical instruments and their many moving parts.

For the new flight instrument CRTs, Collins engineers chose a unique method of display using shadow mask tubes. A shadow mask tube allows a full-color presentation and a bright, crisp picture. Although the principle had been applied to home color television sets, no avionics manufacturer had been able to make a shadow mask tube rugged enough to reliably withstand the vibrations of flight — until the Collins Air Transport Division and its tube supplier developed the technology.

The division was the only avionics manufacturer to develop shadow mask tube technology, but the gamble paid off in 1978 when the Boeing Commercial Airplane Company chose Collins systems for its new-generation airliners.

For the Boeing 757 and 767, two systems using Collins color CRTs were ordered, both representing a radical departure from conventional electromechanical instruments and sparking a revolution in aircraft flight deck design. Instead of intricate moving parts, the electronic flight instruments presented crisp images on a CRT screen. By turning a switch, the pilot could change the type of display on the instrument.

› Principle of shadow mask color cathode ray tube.

BLUE
GREEN
RED
ELECTRON GUN
SHADOW MASK
MASK APERTURE
PHOSPHORS ON GLASS FACE PLATE

> Left: The Boeing 757 and 767 flight decks feature Collins electronic flight instruments, electronic engine indication instruments, digital autopilots and other modern systems.

Boeing photo.

> Right: Boeing 767 and 757 in flight.

Boeing photo.

With these systems, a new era of information access and aircraft system management was opened to the flight crew.

The information was presented in full-color, sunlight-readable displays directly in front of the pilot and co-pilot.

Never before had such a massive amount of data been available for selection by the crew. The technological task facing Collins engineers was to simplify the displayed information in order to reduce cockpit workload and minimize training for pilots learning to fly the new aircraft.

Collins engineers met this challenge by providing flight-critical data in meaningful formats similar to what the pilots were accustomed to seeing on their old instruments. The system could automatically display critical information while in flight. Additional information was available with the push of a button.

The first electronic flight instrument system certified by the FAA was the Collins EFIS for the Boeing 767 in July, 1982.

Digital electronics

Not only was it the first FAA-certified use of CRTs as primary flight instruments, the 767 was also the first completely "digital" aircraft, a technological step just as significant to the aviation industry.

In switching to digital computer-based avionics systems, the aviation industry was removing its 40-year-old "analog" jacket and ordering a new tailor-made "digital" suit.

The contract which led to the first completely digital avionics systems had its origin in the early 1970s.

"We analyzed what we could do with the technology," said Jim Churchill, then vice president and general manager of the Collins Air Transport Division, "and went after the 767 business before Boeing even had a name for it. But we knew they were coming with a new family."

(In March, 1981, Churchill succeeded Rice as president of the Avionics Group, making him Rockwell's senior executive in Cedar Rapids. Rice was appointed vice president of marketing and assistant to the president for Rockwell's Commercial Electronics Operations, and retired in 1983.)

When the division assumed the task of designing the new system, it was more than just a technological challenge. Another hurdle was the industry-wide competition.

> Digital avionics were demonstrated to the world air transport aviation community in April 1980, when the Collins Air Transport Division hosted a three-day symposium in Cedar Rapids. Nearly 200 industry personnel attended, representing airlines, airframe manufacturers and industry organizations.

At the conclusion of the contract awards, the Collins Air Transport Division had won three of the first four major avionics systems for the 767 and 757. They were the flight control system, electronic flight instruments system (EFIS), and the engine indication and crew alerting system (EICAS).

The Boeing 767 entered commercial service with United Air Lines in September, 1982. British Airways and Eastern Airlines introduced the 757 early in 1983. A European consortium of manufacturers — Airbus Industrie — also introduced new digital jetliners featuring many Collins digital units.

The digital concept was adopted by new-generation aircraft in the business aviation and regional airline markets as well.

In the summer of 1981, Collins equipment was selected for the Saab-Fairchild 340, the Short Brothers 360, and Embraer's EMB-120 *Brasilia.* Other turboprop and business jet contracts followed as the Collins General Aviation Division led the way for the business and commuter aviation conversion to digital avionics.

In September, 1982, the first CRT flight instrument for general aviation — a Collins EHSI-74 — was certified aboard a Commander 690 turboprop.

In January, 1983, an entirely new line of digital radios for business, regional airline and military aircraft — Collins Pro Line II — was introduced by the Collins General Aviation Division.

A new product introduced in 1983 by the Collins Air Transport Division was a digital color weather radar for jetliners which could detect turbulence in precipitation — a first in the industry.

An important program for the Collins Government Avionics Division in the early 1980s was an advanced system for the U.S. Air Force that could display friendly and enemy aircraft, navigation checkpoints, hostile surface-to-air missile sites and airfields. The program was called Joint Tactical Information Distribution System (JTIDS).

The portfolio of new products at the Collins Telecommunications Products Division included a communications processor (SELSCAN) that automatically scanned a number of frequencies, determined their availability and quality, and established a confirmed link on the best available channel.

> Left: A Commander 690A turboprop equipped with a four-inch-wide navigation CRT. The EHSI-74 can be seen directly in front of corporate pilot Charles Hall.

> Right: The first all-digital business and commuter-class aircraft was the Saab-Fairchild SF-340, which features Collins avionics as standard equipment.

Saab-Fairchild photo.

> The Collins WXR-700, the world's first digital color weather radar system for airliners, not only displays storm cells and measure rainfall rates, but detects turbulence even in light rain (magenta color on display).

The G-1 — a special lady

› Collins Flight Operations pilot Chuck Hall gives the special G-1 airplane that he'd flown for over 4,500 hours a fond kiss goodbye.

In the spring of 1983, veteran Collins Flight Operations pilot Chuck Hall bade farewell to a very special lady. He did so with a lump in his throat and a pause to reflect upon the many hours — over 4,500 — he had shared with her.

"She's taken me all over the world and never let me down," he recalled. "She's just like an old brood cow — gentle. There's not a mean bone in her body."

She was readily recognizable as the Collins Golfstream I, or G-1 — the 144th twin turboprop to roll off the old Grumman Aircraft Engineering Corp. production line in Bethpage, Long Island, New York.

That was in February 1965. In mid-March of 1983, she was ferried to Columbus, Indiana, for delivery to the Seven-Bar Flying Service — only the second owner in her 18 years.

Throughout the prime of her life, No. 144 covered the skies between Cedar Rapids and Dallas (Addison Field), shuttling employees to and from Collins engineering and manufacturing facilities in the two cities. For a number of years, she performed similar duties on a Dallas-Newport Beach, California, route.

Among her countless passengers were top Collins and Rockwell executives — Arthur A. Collins, W. W. (Bill) Roodhouse, Robert C. Wilson, Donald R. Beall, Martin D. (Skip) Walker, Clare I. Rice, James L. Churchill — who "enjoyed her comfort on a flight tailored to the businessman," said Hall.

During the days of the Apollo lunar landing program, No. 144 frequently was called upon to transport the families of American astronauts from Houston to Cape Canaveral.

The Collins G-1 hosted some well-known entertainers, too. On separate occasions, John Wayne and Bob Hope were given ground demonstrations of Collins systems aboard the aircraft.

Down through the years, the big twin's most important role has been that of a flying testbed and customer demonstrator for major Collins avionics systems designed and developed in Cedar Rapids.

Customer demonstration tours took the G-1 to countries throughout the world — 21 in all — including Canada, Mexico, Great Britain, Scotland, Sweden, France, Belgium, West Germany, Denmark, Italy, Spain, Portugal, Greenland and Iceland. And she appeared for many years at the now-defunct Reading (Pa.) Air Show, and the annual National Business Aircraft Association (NBAA) meeting and convention.

Altogether, No. 144 compiled an impressive list of statistics in her years under the Collins and Rockwell banners, including:

19,800 accident-free flight hours (92 hours per month), a total of 5,702,400 miles — equivalent to nearly 12 round-trip flights to the moon.

8,000 landings, during which the aircraft used 40 sets of tires.

Six major overhauls of the original set of Rolls Royce Dart Mark 529-8X turboprop engines which burned fuel at the rate of 220 gallons per hour and consumed a total of 4,356,000 gallons of kerosene.

She became the first G-1 in Grumman (now Gulfstream Aerospace Corp.) history to reach the 20,000-hour-inspection plateau.

For all of her storied history, No.144 has not been the only G-1 to grace the Collins hangar, but she does represent the lone example of a new G-1 to spend a near full life based in Cedar Rapids.

Over the years, the company has owned five G-1s, but no more than three at any one time.

One of those — No. 163 — was destroyed July 11, 1975, at Addison in one of the first major aviation accidents attributed to wind shear. The quick thinking and intuitive responses of Capt. Dave Selzer and co-pilot Dave Tschudi saved not only their own lives, but seven others aboard the craft. Moreover, all nine emerged from the incident without serious injury.

The good fortune of the passengers and crew was due in part, said Hall, to the structural integrity of "any" airplane built by the 'Grumman Iron Works' — a phrase coined by World War II pilots of LeRoy Grumman's famed 'Cat' series of planes (Wildcat, Hellcat, Tigercat, Bearcat) which survived more than their share of hard landings on such surfaces as aircraft carrier flight decks.

None will be remembered quite like the first — No. 144. From those who most often strapped themselves into her left seat — Harvey Hop, Dick Johnson, Clayt Lander, Selzer, Hall, "Jug" Hiser and Gil Strait — to the thousands of passengers who appreciated her big-cabin comfort and other airliner-like amenities, she may always be known as the 'Queen of the Fleet.'

And she will be missed.

"She was the first big airplane I ever flew," said Hall. "I cut my teeth on her."

With Johnson, pilot Barry Brown and mechanic Randy Weyer, Hall cut his teeth on trans-Atlantic flying, too, ferrying the big turboprop from Cedar Rapids to Toulouse, France, for customer demonstration of new digital air transport avionics in 1979.

Now the role of avionics testbed, customer demonstrator and employee shuttle will fall to a new generation. "She's being replaced by smaller, more fuel-efficient aircraft," explained Hall, who likened the sale to "the shooting of an old horse."

Said he, "She was a peach of an airplane, and a true airplane — one that would tell you when she was going to do something before she did it."

"She will not soon be forgotten." ▪

By the late 1960s, the fleet of Collins aircraft included a Beech Twin and three Grumman Gulfstream I turboprops.

Recession effect

In the spring of 1980, the economic recession, which was ravaging the automobile and housing industries, made its presence felt at the Collins divisions in Cedar Rapids.

In particular, the air transport and general aviation markets were suffering from business slowdowns. Deregulation in 1978 meant major changes in the way the airlines conducted business, and many airline procurement programs were disrupted.

Consumer spending slowed to a crawl because of tight money supplies and high interest rates, which led to smaller corporate profits. Many companies, in turn, postponed or cancelled plans for purchases of business aircraft and the avionics that went with them.

In the early 1980s, employment cutbacks were made in the Cedar Rapids divisions to reduce operating costs as the management team attempted to cope with stagnant economic conditions.

Amid rumors that Collins operations were to be pulled out of the city, management assured employees that the Rockwell International divisions were "deeply committed to Cedar Rapids, which continues to be our center of excellence for engineering and manufacturing."

However, Clare Rice and Jack Cosgrove (CTPD vice president and general manager) each said times ahead would be tough. More cost-cutting measures would be taken to keep the company competitive in its markets.

Those measures included taking a more involved role with employee health care costs. In January, 1981, the company opened its own pharmacy as an added service for employees and families, and to cut the company's health insurance costs.

Through interactive terminals, such as this one operated by Cindy Meihost, Cedar Rapids-based engineers have direct access to the high-speed computing facilities of Rockwell's Information Systems Center, in addition to local processing systems.

› Automated equipment is used to test circuit boards before they are integrated into final products. Mary Taylor is pictured operating the machine.

› Collins Air Transport Division Process Center employs advanced automatic component insertion, wave solder, wash, test and conformal coating techniques to produce multilayer circuit boards.

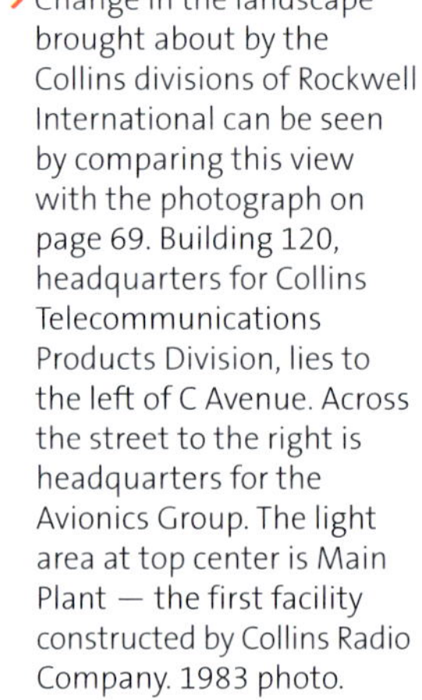

› Change in the landscape brought about by the Collins divisions of Rockwell International can be seen by comparing this view with the photograph on page 69. Building 120, headquarters for Collins Telecommunications Products Division, lies to the left of C Avenue. Across the street to the right is headquarters for the Avionics Group. The light area at top center is Main Plant — the first facility constructed by Collins Radio Company. 1983 photo.

"We are well postured so that if the economy turns, and we can hold our costs in line, we're going to enjoy a tremendous gain and experience prosperity," Churchill said in a 1983 address to Avionics Group managers.

"We've got to be better, just simply better. We need to give more value for the dollar, more product features, and we've got to provide better service. That's where we're trying to spend our money today . . ." ▪

> Arthur Andrew Collins has been called a rare man of greatness. He broke barriers of conventional knowledge and made routine a pattern of scientific advancement.

Arthur A. Collins

Arthur A. Collins, founder of Collins Radio Company, has been called a rare man of greatness, a complex but shy person, who continually had his sights not on the epicenter, but beyond the horizon of technology.

He broke barriers of conventional knowledge and made routine a pattern of scientific advancement. For him and the company he founded and directed for 40 years, work brought many personal rewards, but it was the technical breakthrough and total utilization of technology which provided the greatest motivation.

Collins attended Cedar Rapids public schools, Amherst College in Amherst, Massachusetts in 1927, Coe College in Cedar Rapids for special courses, and the University of Iowa for advanced studies in physics.

He received an honorary doctor of science degree from Coe College in 1954, an honorary doctorate from the Polytechnic Institute of Brooklyn in 1968, an honorary doctor of engineering degree from Southern Methodist University in 1970, and an honorary doctor of science degree from Mount Mercy College in Cedar Rapids in 1974.

From 1945 through 1951, Mr. Collins served as a director of Coe College. He was a director of the Graduate Research Center of the Southwest, Dallas, from 1962 to 1969, and served on the board of directors of the Herbert Hoover Foundation.

He was a member of the International Sponsors Committee of the Robert Hutchings Goddard Library Program, Clark University, Worcester, Massachusetts.

Mr. Collins belonged to the Institute of Electrical and Electronic Engineers, the Navy League of the United States, the American Ordnance Association, the Armed Forces Communications and Electronics Association, and was a member of the Cedar Valley Amateur Radio Club.

He received the Secretary of the Navy's Distinguished Public Service Award Citation in 1962, the Iowa Broadcasters Association Distinguished Service Award in 1966, was elected to the National Academy of Engineering in 1968, received the Armed Forces Communications and Electronics Association's David Sarnoff Award in 1979, and the Electronics Industries Association's Medal of Honor in 1980.

After leaving Collins Radio in 1972, he formed a new firm, Arthur A. Collins, Inc., based in Dallas, to carry out systems engineering studies in the communications and computer fields.

He was married to Mary Margaret (Meis), and was formerly married to Margaret Van Dyke, who died in 1955. He has four children.

The man who started it all and accomplished so much, Arthur A. Collins, died on February 25, 1987, at the age of 77. In his 2002 biography of Collins — "Arthur Collins, Radio Wizard" — Ben W. Stearns quoted him as telling a long-time colleague just the day before, "There are still so many things I need to do." ▪

The Cold War years wind down

Jim Churchill's view that the Avionics Group was positioned for prosperity was well-founded. By 1984, nearly 50 airlines had placed orders for the Collins WXR-700 Digital Color Weather Radar, and more inroads were made in digital avionics for regional-service aircraft. An agreement with Beech Aircraft meant its new Starship 1 advanced-technology business aircraft would feature a complete Collins digital cockpit system. Overall, makers of more than 30 new business aircraft had selected Collins Electronic Flight Instrument Systems (EFIS).

The commercial aviation markets in general remained depressed, but all-digital Collins avionics with color CRT displays were still being placed on Boeing's then new 757 and 767 aircraft.

‹ The Boeing 757 (foreground) featured all-digital Collins Avionics with color CRT displays, as did the Boeing 767.

› More than 2,000 Collins WXR-700 weather radar systems had been installed by the end of the 1980s.

President of the Avionics Group from 1981 until his retirement in 1990, Churchill later recalled that, just as winning the L-1011 business earlier had given the Avionics Group valuable credibility that was necessary for it to compete successfully for the 757 and 767 positions, those Boeing successes paved the way for inroads into the burgeoning business aviation market.

"We were able to generate enough cash and profits to allow us to make the kind of investments it took to broaden not only our product line, but our systems development," he recalled. "We had always concentrated on the heavier air frames and the military side. Once we got serious about business jets and general aviation at large, Pro Line development was a key factor."

Getting serious meant developing an integrated systems approach geared specifically for corporate and regional airline markets, which were addressed by new products as well as refinements in Pro Line avionics. The Collins Avionics

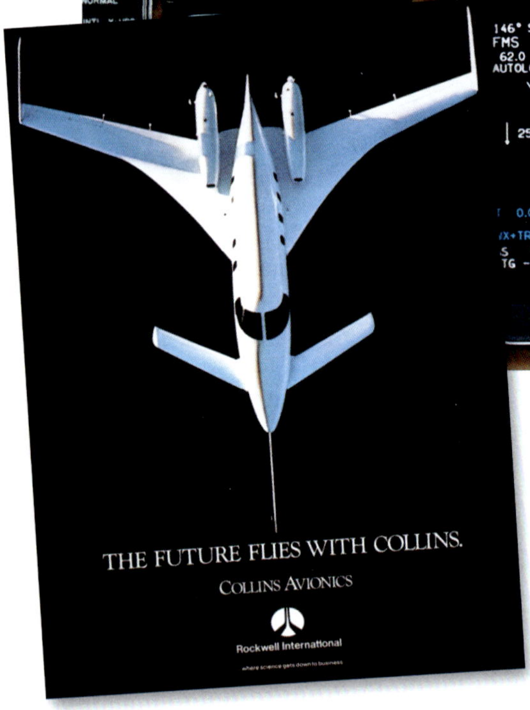

Group's General Aviation Division, for example, unveiled the first fully digital autopilot and auto-throttle systems for business jets in March 1984, attracting numerous aviation-industry journalists to a press conference in Cedar Rapids. The APS-85 and APS-95 autopilot systems both used digital inputs from the Collins AHS-85 attitude heading reference system. The new ATS-85 auto-throttle system was designed to complement both autopilots and to operate up to four throttles smoothly and efficiently for optimum use of fuel in automatic landing situations as well as in climbing and descending modes.

Over the next several years, Collins avionics became standard or standard options on a broad range of corporate and regional aircraft. The Pro Line 4 integrated avionics system alone was selected for nine new business and regional airline aircraft in that period.

› The cockpit of the unique Beech Starship 1 business jet featured a Collins Pro Line 4 advanced avionics system. It was the first business jet with an all-glass cockpit.

› The Collins APS-85 was the first fully digital autopilot system for business jets.

Military contracts fuel growth

While the commercial business would see its own boom later in the 1980s, military contracts were the biggest source of growth through most of the decade. With Ronald Reagan as president and a philosophy of "peace through strength," U.S. defense spending increased by 35 percent from 1981 to 1989. That military buildup was widely seen as a major contributing factor to the fall of the Berlin Wall in 1989 and the eventual disintegration of the U.S.S.R.

As the defense industry would learn later, those events signaled a subsequent downturn in military spending. In the meantime, however, the corporation and the Collins businesses continued to make their mark in notable new military programs.

Collins Defense Communications (CDC) was focused primarily on naval and land forces. The primary market for the Collins Government Avionics Division (CGAD), as part of the separate Avionics Group, was in airborne electronics.

Rockwell International had a major role in the Air Force's B-1B bomber program, in which the Collins businesses also participated. Built in Rockwell's North American Aircraft Operations complex in Palmdale, California, the B-1B carried six Collins avionics and telecommunications systems. Three of them had been designated Air Force standards: the ARN-118(V) tactical air navigation (TACAN) transceiver, the ARC-190(V) HF communications system and the ARC-171(V) UHF transceiver. The Collins TACAN system was operational on Army and Navy aircraft and in the fleets of 35 other countries. The B-1B aircraft also were equipped with the Collins ARN-108 instrument landing system, two 562A-12A flight director computer monitors and CTS-81 automatic test equipment.

Collins' technological expertise had made aviation and GPS history in May 1983 when Collins Flight Operations director and pilot Chuck Hall and his crew made the first trans-Atlantic crossing using satellite navigation. On a flight originating in Cedar Rapids and ending in Paris, three world aviation records were established. That flight set the stage for numerous Collins GPS program wins and production contracts.

CGAD had begun delivering Navstar GPS user equipment in 1983 as it competed for a Defense Department production contract. When the contract was awarded the following year, Collins employees hung banners in the atrium lobby of Building 124, congratulating the GPS team and announcing, "The FROG has LANDED." They had devised a Monopoly-type board game using frogs to represent the user-equipment competitors. The Collins frog had won the game.

› Among six Collins avionics and telecommunications systems on the B-1B Lancer bomber (foreground, top) were the ARC-190 communications system (above) and the ARC-171 transceiver (below).

> The successful Navstar GPS system helped revolutionize navigation technology for both military and civilian use.

Satellites show the way to Paris

On May 22, 1983, a Rockwell Sabreliner dubbed "Navstar 1" took off in the dark from Cedar Rapids. The stars were out, but using them to navigate, of course, was a thing of the past. The way to Paris, with four stops on the way, would be guided by five Navstar satellites and, on board, both military and civilian GPS sets. With Chuck Hall, Collins Flight Operations director, the crew included Chief Pilot Dave Selzer; navigator Loren DeGroot, director of Avionics Systems for the Collins Government Avionics Division (CGAD); and David Van Dusseldorp, the division's manager of GPS Manpack and Vehicular Engineering.

The July 1983 issue of *Rockwell News* reported that limited satellite coverage meant the crew had to plan on stops in Burlington, Vermont; Gander, Newfoundland; Reykjavik, Iceland; and Luton Airport, near London. Each leg of the flight was flown at night when the satellites were in the necessary position. Navstar 1's flying time was 11 hours and 15 minutes spread over the course of just over three and a half days. After 4,228 nautical miles, the plane rolled to a stop on May 26 just 7.5 meters — about 26 feet — from its target parking spot at Le Bourget Airport. Crew members were greeted with champagne, handshakes and congratulations.

Hall was quoted as saying, "The equipment performed flawlessly. But it took a maximum effort by the whole crew to make the flight a success."

When asked about the significance of the flight, DeGroot said, "We proved that Navstar GPS is here today and that it could be flown in today's air traffic environment. We cannot yet comprehend the impact of what navigation accuracy within several meters and highly precise velocity and time information will mean to us in the future."

The historic flight resulted in four major aviation awards from the National Aeronautic Association and the Federation Aeronautique Internationale. ▪

> The record-setting crew and "Navstar 1" Rockwell Sabreliner were guided on their way from Cedar Rapids to Paris by a fledgling global-positioning system.

CGAD delivered its first preproduction receiver/transmitter for the JTIDS Class 2 terminal program in late 1983. The division designed and developed the R/T to be combined with lead contractor Singer-Kearfott's equipment to form the Class 2 Time Division Multiple Access (TDMA) terminal. The terminals were designed for Air Force fighter and Army ground tactical applications allowing exchange of information over a single communication link.

CGAD also participated in development of data links for transferring target information between scout and attack aircraft. In May 1984, officials turned over two U.S. Army OH-58 helicopters to 9th Infantry representatives in ceremonies at Collins Flight Operations in Cedar Rapids, where employees had installed a new automatic target handoff system (ATHS) and new avionics system. The two helicopters were part of a program, called SCAT-58, to provide fully integrated avionics as part of their scout/attack role for the Army's rapid deployment force. The avionics were controlled by the Collins CMS-80 avionics management system, including the ATHS, communications, navigation, surveillance and fuel management. Other Collins avionics aboard the helicopters included the ARC-186 VHF and ARC-174 HF transceivers.

› Top: The U.S. Army's OH-58 helicopter (shown in the Middle East) featured an automated target handoff system and a new avionics system.

› Above: The Collins Government Avionics Division developed TDMA terminal equipment for U.S. Army and Air Force applications as part of the JTIDS program.

Evolving, growing, delivering

One of a number of reorganization moves took place late in 1984 when Jack Cosgrove was named president of Collins Defense Communications, headquartered in Cedar Rapids. The new organization brought together the previously separate Cedar Rapids-based Collins Defense Communications Division and the Collins Communications Systems Division, which had been headquartered in Dallas.

> Jack Cosgrove was named Collins Defense Communications president in 1984.

CDC employed more than 6,000 people at the time. The new CDC configuration included the Advanced Communications and Countermeasures Division (ACCD) in Cedar Rapids. It also included two divisions with headquarters in Dallas: the Mobile Subscriber Equipment Division and the High Frequency Communications Division (HFCD). HFCD later became known as the Strategic and Tactical Communications Division (STCD) in recognition that its business no longer focused strictly on the HF spectrum.

Both ACCD and STCD were dissolved four years later when CDC was realigned again and their functional staff organizations were centralized. Engineering, logistics, human resources and finance functions were consolidated in the process.

"This will allow CDC to focus its resources on program execution and performance," Cosgrove said at the time.

CDC had celebrated the grand opening of its High Technology Manufacturing Building in Cedar Rapids in July 1985. The 80,000-square-foot, flexible manufacturing facility, dubbed Building 166, was added at the Main Plant complex at 32nd Street Northeast and Eastern Avenue. It was touted as "one of the most advanced manufacturing plants of its kind in the defense electronics industry." A *Rockwell News* article about the building noted that it would use new bar-code technology to manage material handling and track the progress of parts through the manufacturing process. Many other activities were automated, as well, as the plant made extensive use of computer-aided manufacturing equipment.

Also in 1985, CDC won a $256 million Air Force contract, as a team with Ball Aerospace and subcontractor to Raytheon, to develop Milstar communication terminals. Most of that work was to be done at CDC's facility in Santa Ana, California. CDC had also just won a $48 million Air Force contract for full-scale development of VLF Miniature Receive Terminals (MRTs) to be installed on FB-111, B-52 and B-1B bombers. Design and production would take place in Santa Ana, Salt Lake City and Richardson, Texas, facilities. The MRT equipment was part of a system designed to provide a survivable communication link between U.S. command and strategic bomber forces.

Good news and celebrations earlier in the year were followed late in 1985 by a setback for Rockwell International. The corporation entered a plea of guilty in a Dallas federal court to charges that six CDC employees in Richardson, Texas, had submitted false time cards for government contract work three years earlier. The company ultimately paid more than $1 million in fines and settlement costs and was suspended for several weeks from seeking government contracts.

> Military supply personnel on a tour of Collins facilities.

The Avionics Group's CGAD continued its GPS successes with the Air Force announcement in 1985 of an initial purchase of 6,100 Navstar GPS user equipment sets for use by the Air Force, Army and Navy. The Navstar system was touted as "the most precise radio navigation system ever developed." It was to provide airborne, shipboard and land-based users with position information accurate to 16 meters in three dimensions; correct time to within a billionth of a second; and speed data within 0.1 meters per second. Six Navstar satellites, produced by Rockwell's Satellite Systems Division in Seal Beach, California, were already orbiting the Earth 11,000 miles up.

CGAD had spent 10 years working on the Navstar program and had been in a full-scale design and development competition with Magnavox Advanced Products and Systems Company. The Collins division won the Phase III production contract after two years of field testing by the Air Force on several platforms, including submarines, an aircraft carrier, tactical fighter aircraft, strategic bombers, helicopters, jeeps and tanks. A "manpack" unit, which could be carried on a soldier's back or mounted in a vehicle, also was evaluated.

Plans were under way to deliver on the division's GPS orders through a new automated plant being built in Coralville, Iowa. The GPS business would later increase further through expansion into railroad, maritime and even agricultural applications.

The Paris Air Show in 1985 featured a GPS demonstration with a specially equipped Navstar van — the Collins Navstar Shuttle. The van featured a video map display showing the vehicle's position as determined by the onboard Collins GPS receiver and signals from orbiting Navstar satellites. A Collins GPS manpack receiver also was demonstrated daily at the show, where numerous other Collins avionics and telecommunications products were displayed. (Among them were some notable commercial systems, including the EFIS introduced in Boeing's 757 and 767, which featured the first "common cockpit" configuration. McDonnell Douglas, Airbus, Embraer and SAAB-Fairchild also had installed new Collins avionics systems.)

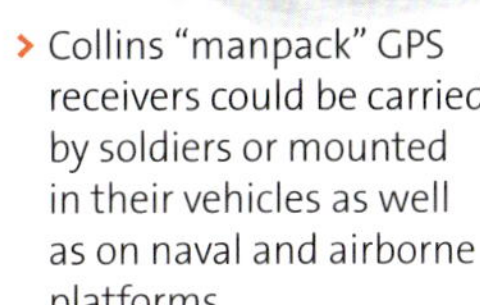

> Collins "manpack" GPS receivers could be carried by soldiers or mounted in their vehicles as well as on naval and airborne platforms.

> The Boeing 757 and 767 cockpits featured Collins Electronic Flight Instrument Systems.

› Herm Reininga, vice president, Operations, displays the ARC-190 Bird shirt he was given to mark the completion of the 5,000th ARC-190 radio.

ARC-190 and the "Bird"

When the landmark 5,000th ARC-190 came off the production line in May 1988, Cedar Rapids employees celebrated with refreshments and speeches and presented CDC President Jack Cosgrove and Operations V.P. Herm Reininga with special "Bird" T-shirts. As a *Rockwell News* article explained, the name came from the B-1 Research and Development Program under which the radio was developed. "In condensed form, that reads 'B-1R&D' or 'B1RD.' Before long, the '1' became an 'I' and the code name 'BIRD' was born." More than 7,000 Bird radios had been ordered when the special shirts were made. The radios had brought CDC some $300 million worth of business over 10 years. ▪

Early 1986 brought news of a $34.4 million contract for CDC to design, develop and produce ARC-190 HF radios and support equipment. It was believed to be the largest single-product award that CDC had received by that point. The solid-state, 400-watt airborne communications system was to be produced in CDC's Cedar Rapids and Toronto facilities for use on 14 different Air Force reconnaissance, cargo and fighter planes and helicopters. The ARC-190 also was sold to the U.S. Navy and Coast Guard as well as to international customers.

The landmark ARC-190 contract was eclipsed in December 1986 by the news that CDC's Advanced Communications and Countermeasures Division had won two contracts totaling more than $46 million for development and production of the ARC-182 UHF/VHF radio system. The ARC-182, which by then had become the standard radio for all U.S. Navy aircraft, was produced in the High Technology Manufacturing Center in Cedar Rapids with extensive use of robotics and surface-mounted component technology. One of the contracts was for development of electronic counter countermeasure (ECCM) capability for the ARC-182. The 5,000th ARC-182 came off the production line in May 1987.

CDC also was awarded two multimillion-dollar contracts over the next several months to provide HF-190 radios for the Gulfstream 4 business jet. The HF-190, manufactured in Cedar Rapids, was the commercial version of the ARC-190 HF transceiver and also was sold internationally for aircraft used by foreign heads of state. The radio's design allowed users to continually transmit data with computer modems without overheating the power amplifier, which was a problem with other HF radios designed for business aircraft.

In the mid-1980s, Rockwell International and the Collins businesses were working to find a new balance between military and commercial, and between domestic and international business. One example was that the primarily government-focused CDC in 1986 expanded its marketing of the innovative HF-9000 radio to Europe's commercial aircraft manufacturers.

The lightweight, rapid-tuning HF-9000 radio was described as ideal for both military and commercial markets. It featured three microprocessors and fiber optics control. Defense users were expected to benefit from using a lower-cost, commercially available product while commercial users were attracted by the

› The HF-9000 system used by both military and commercial customers consisted of a compact control unit, receiver/transmitter and an automatic antenna coupler.

military-grade reliability. Canadair Corp. had already ordered HF-9000 shipsets for its Challenger 601 business jets.

On the other side of the world, CGAD landed a $53 million contract for advanced avionics for a Royal Australian Navy helicopter used in antisubmarine warfare and surface surveillance. Other Collins equipment also was making news well outside the United States — for helping a pilot fly over the North Pole and helping a mountaineering engineer reach the summit of France's Mont Blanc. Dr. Millard Harmon used a Collins Navcore™ I GPS receiver installed in a Beech 36 Bonanza aircraft to navigate over the pole on August 8, 1986. The same day, Jean-Marie Blot climbed Mont Blanc, in the Alps, carrying a portable Navcore™ I GPS unit.

GPS and other programs grow

Following a Critical Design Review in Cedar Rapids in 1987 — which resulted in approval of the Collins five-channel GPS receiver, the standard control/display unit, remote console unit and related antenna electronics — CGAD was approved for continued production and the government exercised its first production option. The division had been awarded GPS production contracts valued at more than $150 million, and much more would follow.

The first production-model GPS user set was presented, ahead of schedule, to the Air Force in a ceremony in September 1987 by Robert Marovich, CGAD vice president and general manager. *Rockwell News* reported that automation had reduced the price of GPS equipment to half of the original cost-per-unit goal. The initial GPS set was to be installed on Navy SH-60 helicopters. Overall, some 22,000 Air Force, Army and Navy aircraft were scheduled to be fitted with GPS units.

An Air Force general predicted an "explosion" of GPS use, primarily in the form of manpack sets, which weighed about 17 pounds and used signals from four satellites to determine position. About 1,400 manpack GPS units were delivered over the next six years. The actual explosion would come later in the form of much smaller, more easily transported units.

A few weeks before delivering that first GPS user set to the Air Force, the Avionics Group had begun operating a demonstration satellite-based communications system using an antenna about the size of a common nail. The Land-Mobile Satellite System (LMSS) used satellites to relay data between an office and a vehicle on the road. It also automatically reported the vehicle's position using GPS. The design, assembly and testing were completed within 10 months.

› Some of the first Collins GPS user equipment sets were installed on U.S. Navy helicopters like the SH-60 Seahawk.

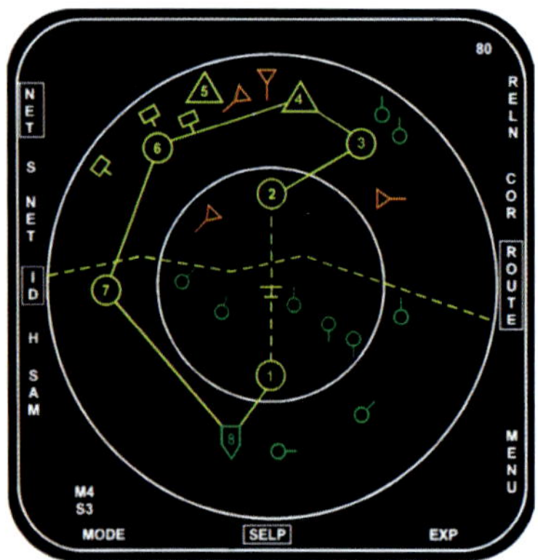

> The JTIDS tactical display indicated the relative positions of friendly and enemy combatants.

CDC scored an important win in 1987 after teaming up with Marconi Electronic Systems to bid on a major communications project for the U.S. Navy. The High-Frequency Anti-Jam (HFAJ) program was designed to provide naval forces with survivable HF communications that would be immune to electronic jamming. The contract had an initial value of $42.2 million and much greater potential value. The Navy contract called for development of both the HFAJ system and the Link Eleven Improvement Program (LEIP) system.

Such joint ventures were becoming increasingly common, particularly in dealings with the Department of Defense, which was beginning to require them to ensure that more than one provider could ultimately deliver mission-critical technology. Jack Cosgrove, who headed CDC as president from 1981 to 1990, noted that Marconi was a major competitor much of the time. He added that teaming up with competitors had been considered "blasphemy" until the mid-1980s.

The Avionics Group's CGAD also teamed with Singer-Kearfott in a leader-follower arrangement to win an Air Force program to develop and produce data links for F-15 aircraft. That program, the Joint Tactical Information Distribution System (JTIDS), would loom large in the history of the Collins business in more ways than one.

CGAD received two special honors in 1988. It was named Air Force Programs Contractor of the Year for GPS-related achievements in value engineering that were expected to save the Air Force millions of dollars. It also received the 1988 Outstanding Technical Achievement Award from the Defense Advanced Research Projects Agency (DARPA) for work on miniaturizing GPS receivers. DARPA's Aerospace and Strategic Technology Office cited CGAD for its development and demonstration of the first Monolithic Microwave Integrated Circuit (MMIC) chip, which was designed for the GPS program.

The MMIC chip was developed under a DARPA contract to produce a handheld GPS receiver that could be no larger than 100 cubic centimeters. The Collins division had demonstrated that the use of MMIC and Very Large Scale Integration (VLSI)

> Development of the MMIC chip made it possible to produce small, handheld GPS receivers.

DARPA photo.

technologies eventually could lead to enormous reductions in the size and cost of GPS equipment and other electronics for both military and commercial use.

CGAD was also in its first full year of Navstar GPS production in 1988. In addition, it had won orders from the Department of Defense and foreign militaries for upgrading a variety of aircraft with integrated cockpit management systems.

CGAD won yet another GPS contract in May 1988, this time to provide Navstar user equipment for the U.S. Navy's P-3 Orion aircraft, its primary anti-submarine warfare platform. The Navy was in the process of upgrading its P-3 avionics in response to increasingly quiet and more dangerous new Soviet submarines. The next month, the division announced it also had won a $41 million contract to upgrade AH-64A Apache attack helicopters by integrating cockpit management systems. Dick Marett, the program manager, said the Apache program solidified the company's credibility with the Army as a major player and a technically competent system integrator. He was quoted as saying, "This is the first time the Army has awarded integration of a major front-line system to a company other than an airframe manufacturer."

Key elements of the AH-64A system included the Collins Airborne Target Handover System (ATHS), dual control-display units and a data transfer system. CGAD subcontracted the development of system kits and installation work to its fellow Collins defense contractor CDC, which was opening its new Aircraft Modification Center in Shreveport, Louisiana. That new facility would be used for helicopter work as well as upgrading of much larger fixed-wing aircraft. The 108,000-square-foot hangar with 67-foot-high doors was designed to accommodate planes as large as the Boeing 747 and the military's wide-body C-5. It opened in May 1988.

> The Collins Government Avionics Division provided Navstar GPS user equipment for the P-3 Orion as the U.S. Navy upgraded its anti-submarine warfare capabilities.

> CGAD's selection to provide an integrated avionics system for the AH-64A Apache attack helicopter was a first for the U.S. Army. CGAD and CDC both worked on the project.

> A lightweight Collins TACAN system was chosen to replace an earlier system used by NASA's Space Shuttle program.

NASA photo.

Later in 1988, CGAD was awarded a contract to provide Integrated Communication Navigation Identification Sets (ICNIS) in three versions of the Air Force's F-111. The following year, the division was selected to supply GPS user equipment for the Air Force F-111D and F-111E fighters under a modification program dubbed "Pacer Strike." Among other upgrades, CGAD supplied ICNIS equipment for interface, control and display of the fighters' cockpit avionics.

The U.S. Coast Guard also selected Collins avionics for an upgrade to its fleet of HH-65A short-range recovery helicopters, which required precise, all-weather navigation capability. The integrated GPS/avionics system was designed to enable the GPS receiver to work as a navigation computer in the event of a primary mission computer failure. It gave both the pilot and co-pilot individual control display units that provided access to GPS data and all navigation, communication and other avionics information. The installation work was done at the new modification center in Shreveport.

The Space Shuttle program was largely the business of Rockwell International's Aerospace Division, but Collins avionics did play a role in space exploration even after the company's previous Apollo involvement in transmitting historic data, voice and video communications from the moon back to Earth. In 1989, Collins technology was tapped for an update to the Space Shuttle when CGAD was chosen to provide a lighter-weight replacement for the shuttle's existing TACAN system, which guides the craft during landing.

> Integrated GPS/avionics systems for Coast Guard helicopters were installed at the Collins modification center in Shreveport, La.

While there were significant successes in the booming military electronics industry of the 1980s, there were also some costly lessons to be learned. Jack Cosgrove recalled that big military programs sometimes brought big, expensive headaches. A Navy contract for High-Power Transmit Sets, for example, proved to be a losing proposition. The division underestimated its costs on the fixed-price program for state-of-the-art VLF transmitters. The lessons learned were reinforced in a later contract to provide ground terminals for the Pentagon's Milstar satellite communication system as a subcontractor to Raytheon.

"Both programs were big losses," Cosgrove recalled. "That taught us a lot of valuable lessons but it was very expensive. It was painful. From that, we learned how not to go do something."

The experiences led to the establishment of a new value proposition that would require good answers to three questions before new business would be pursued: Is it real? Can we win? Is it worth it?

> The Boeing 737-300 featured Collins electronic flight instrumentation.

Boeing photo.

Commercial gains

While data links technology was increasingly important to the military, the Avionics Group's Collins Air Transport Division in 1984 was working on a civilian data link system. Known as the DL-700 system, it was being developed to improve message processing between ground facilities and commercial airliners.

The General Aviation Division also was breaking new ground. In October 1984, it introduced a new digital color weather radar system — the WXR-350 — designed to work with electronic flight instrument systems. Part of the Pro Line II family, the radar featured a new path-attenuation-correction alert feature that identified potentially dangerous cells hidden behind heavy rainfall. It also provided ground mapping capability.

Two months later, the division announced the FAA certification of the "world's first digital business jet" — a British Aerospace BAe 800 equipped with "a full complement of the most advanced Collins digital avionics systems." The Collins equipment included an APS-85 digital autopilot system, an advanced EFIS-85 system, ADS-92 air data system, AHS-85 attitude heading reference system, and Pro Line II digital radios.

Soon after, in 1985, came the announcement that the Air Transport Division would provide its Electronic Flight Instrumentation System for the Boeing 737-300. The system was a derivative of the EFIS-700 provided earlier for the Boeing 757 and 767. It featured a full-color CRT display system for primary instrumentation.

The company delivered the first display and maintenance system for the Boeing 747-400 in August 1986, giving the Avionics Group significant display systems on all of Boeing's major commercial aircraft: the 737, 757 and 767 in addition to the 747. The avionics update replaced the 747's electromechanical instruments beginning in 1987 and significantly reduced the number of instruments in the cockpit, which contained six interchangeable Collins DU-7000 CRT display units. A Boeing executive said at the time that the system's flexibility would "so significantly reduce the workload on the flight deck that a two-man crew will replace the three people it now takes to operate the aircraft."

Software for the system was designed in the Ada computer language by a team of engineers working in Cedar Rapids, Melbourne, and Downey, California. It was the first time the language had been used in a Boeing production program.

The 747 system also was notable in that the new Central Maintenance Computer System could monitor up to 65 subsystems and perform tests of all the aircraft's

> The BAe 800, which featured the most advanced Collins avionics available, was the first "digital business jet."

> When the Boeing 747-400 entered service, Collins avionics and communications systems figured prominently, and Collins' position continued to grow with the 747's success.

Top: Boeing photo.

electrical, electromechanical and avionics subsystems. The 747-400 entered service in 1989 equipped with all-new Collins automatic flight control and cockpit display systems and maintenance computers in addition to communications, navigation, identification and radar systems.

Orders for new commercial jetliners reached an industry high in 1988 and nearly all of the free world's major airlines were equipped with Collins avionics. Collins' position on the Boeing 747 continued to grow and the company was gaining ground overseas, which accounted for more than half of its commercial aviation sales.

The Collins Air Transport Division reached a major milestone in production of its WXR-700 weather radar in 1988 by delivering its 2,000th system. The WXR-700 had been installed aboard some 1,500 aircraft operated by 76 airlines worldwide. They had amassed more than 15 million flight hours with the system, which was the first solid-state, all-digital radar system for the airline industry. Ground-based applications of the system were in use by 10 civilian airports and military air bases and more than 45 television stations. One of the first TV stations to display storm-related turbulence with a Collins-designed Doppler weather radar system was KGAN, Channel 2, in Cedar Rapids. (The first Collins weather radar for an airport had been installed in 1985 at the Beaver County Airport in Pennsylvania.)

The General Aviation Division made news by becoming the first avionics producer to deliver turbulence-detecting radar in the business aircraft market when it introduced the TWR-850 Doppler weather radar. The TWR-850 operated at 24 watts, a small fraction of the power previous radars required, thanks to its solid-state design and the sensitive receiver's use of gallium arsenide chips rather than conventional silicon. The division also was celebrating certification of the Beech Starship, whose cockpit — with a Collins Pro Line 4 advanced avionics system — represented a first in business aircraft design: The airframe and avionics were designed concurrently.

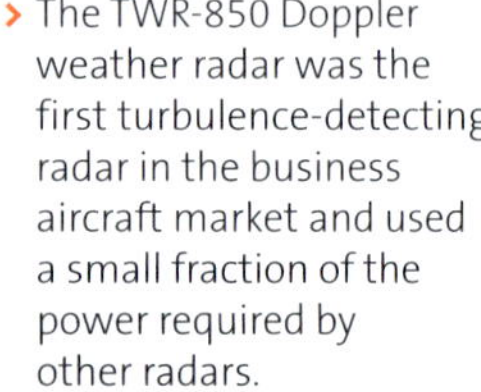

> The TWR-850 Doppler weather radar was the first turbulence-detecting radar in the business aircraft market and used a small fraction of the power required by other radars.

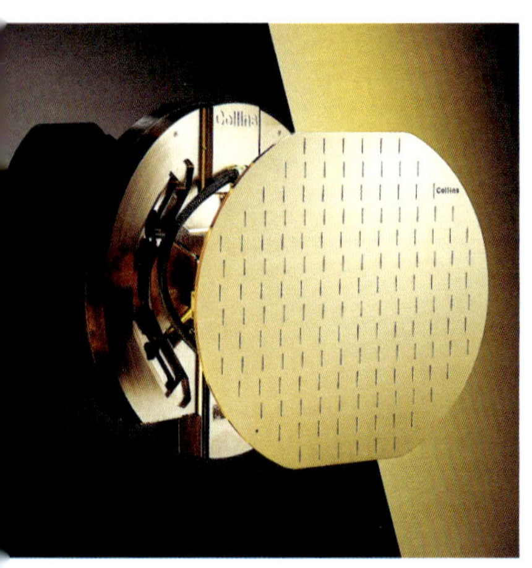

> The flight deck of the Canadair regional jet featured the large-format displays of the Collins Pro Line 4 advanced avionics system.

The Starship program was described as "the largest single undertaking the division has ever managed."

By 1989, the Pro Line 4 system also had been selected for a new Canadair regional jet — the first regional airliner to go into commercial service equipped with the advanced integrated avionics system — and also for the Saab 2000 regional aircraft, the Beechjet 400A and Embraer's CBA-123. The Pro Line 4 system also was adopted for a regional jet under development in Indonesia.

Collins engineers at the time had also begun working on the next-generation of Traffic-alert Collision Avoidance Systems (TCAS) in anticipation of strict new FAA regulations. The Collins TCAS II was being flight-tested for proof-of-concept certification. TCAS was seen as a natural and growing market for the company's technical expertise, but Collins was a latecomer to the business and competition proved fierce. Several airlines chose competitors' systems early on, but both Air Canada and Pan Am chose the Collins system for their fleets in 1989, positioning the company as a legitimate TCAS provider. Even bigger success would come soon after, when United Airlines selected the Collins TCAS II for installation on more than 450 of its aircraft and established it as the TCAS market leader. United's choice stunned the industry, as it had widely been expected to select a rival Bendix system. By the end of 1989, more than a dozen major air carriers had ordered Collins TCAS II systems.

> The Collins Traffic-alert Collision Avoidance System gave pilots a vivid display indicating the relative positions of other aircraft in the area, including their distance and altitude.

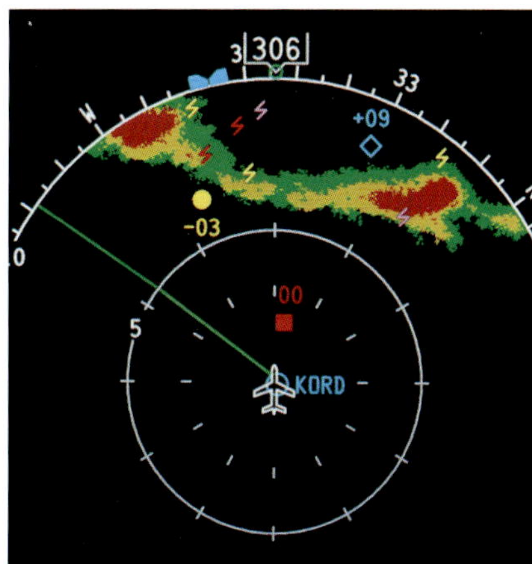

Industry changes, cultural changes

Winning new business that was worth winning depended increasingly on automation, software development — trimming time, weight and size — and finding other ways to drive down costs. An industry and company that were largely oriented to making hardware — bulky radios, for example, with through-hole-mounted transistors and components — were having to adapt to a software-dominated world that was peopled with college graduates who didn't look or act like the stereotypical engineer in white shirt and narrow tie.

People had to learn to think differently on several levels, on both the commercial and military sides of the business.

"The change from hardware to software is probably the most dramatic change that happened in that period of time," said Cosgrove. "It seems like it changed overnight. We went from conventional parts to surface-mounted parts to integrated circuits, and along that same time we went from 80 or 90 percent hardware to 10 or 20 percent hardware."

Software engineers weren't as likely as others to show up promptly for work at 7:30 a.m., Cosgrove recalled, and they often looked different, too.

"They did pretty good work, but they were unconventional," he said. "We had guys wearing ties and they'd come in with their Hawaiian shirts hanging out and wearing shorts and sandals. It was quite a cultural shock for a while." The cultures eventually merged into a thoroughly professional unit, but that wouldn't be the last culture shock.

The "Computer Age" had arrived and businesses had to compete for people with new skills. New ways had to be devised to entice them to come to Cedar Rapids and stay. Manufacturing was changing. The world was changing, and people would have to change, too. Eventually, hardware-oriented engineers and software developers would get comfortable with each other and a new and different culture would emerge. Engineers would need to integrate what they did more closely with factories. For the Collins businesses to stay competitive, leaders would need to emerge and be trained at all levels.

Other workforce-related changes were taking place. The addition of a pharmacy in Cedar Rapids had been one step toward cutting health-care costs for employees as well as for the company. Later in the 1980s, establishment of a local Health Care Council with other Cedar Rapids businesses and creation of a preferred provider organization of local doctors and hospitals "blew the socks off of health care costs" — at least for a time — in the words of Dick Johnson, who was V.P. of Human Resources for Collins Commercial Avionics.

> The company-owned pharmacy helped control health-care costs and made prescriptions more affordable for employees.

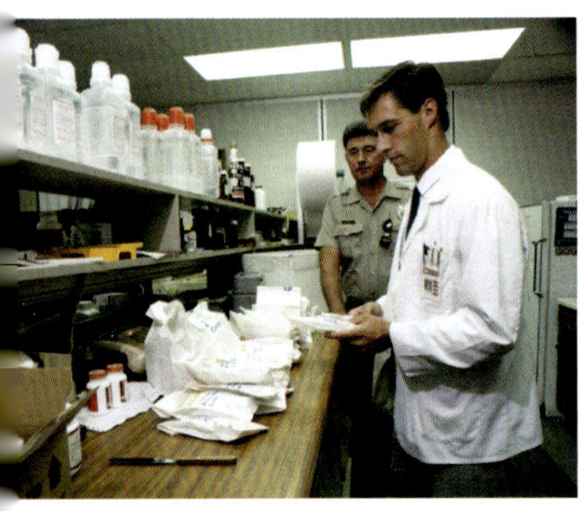

Two notable employee-relations landmarks were put in place in the mid-1980s. One was prompted by a question that Jim Churchill posed to Johnson, who recalled, "Jim said to me, 'How do these young women who work for us on the line and out in the factory make it when they've got one or two kids, and maybe they're single mothers or divorced, what do they do with their kids?'"

Recognizing that caring for children and holding down a job could be challenging, he asked Johnson to look into setting up a child care center. Building 154, which had once housed the Collins Switching Systems Division, was available. That became the original Child Development Center, which opened in April 1986 as the first corporate child care center in Cedar Rapids. It started with a staff of eight caring for 30 children. An addition to that building became the company's Recreation Center, which opened

Recreation and child care centers open

The Recreation and Child Development Centers were dedicated April 12, 1986. That weekend, according to a *Rockwell News* report, more than 10,000 employees and family members toured the facilities at Collins Road and Rockwell Drive Northeast in Cedar Rapids. Avionics Group Personnel Resources Manager Dick Fredericks noted that the Child Development Center was the largest company-owned facility of its kind in the United States. Two years after the center opened, the Congressional Caucus for Women's Issues praised the company's efforts with special recognition through its Child Care Challenge. The Child Development Center would eventually outgrow its original home and move into new quarters in 2009. ▪

at the same time. By opening day, more than 1,500 employees had signed up to use the recreation facilities. A basic individual membership was $11 per month.

Churchill credited Johnson and others with identifying the need to provide child care and other services. "It was a very responsible thing to do," he said, and providing a convenient child care center also showed employees that the corporation cared about their families.

"It gave them a chance to worry about their children but not be overly concerned and still be able to do their work," he recalled. "More than anything I saw it as a benefit that was needed for the existing employee base." Recruiting wasn't the motivation initially, he noted, although the recreation and child care services did add to the company's appeal as a good place to work.

Johnson remembered the 1980s as times of innovation in employee relations as well as in product development. "We were 10 years ahead of our time," he said. The pharmacy, child care and recreation center all became important incentives for attracting and retaining talented employees.

Those employees also have been forward-thinking in sharing their expertise and encouraging young people to explore science and technology.

Johnson noted that the Partnerships with Education program originated with a group of employees led by George Benning, who asked him, "What do you think of the idea of us getting more involved with some of the schools in the community? We could have some of our people go into the schools and teach how rockets work and so forth."

The school principals who were approached about the idea were skeptical at first. Johnson recalled one of them asking, "What's in it for you? Why are you doing this?"

The response, Johnson said, was, "We don't have any agenda. We just think that it's the right thing to do from a community standpoint and we have a lot of our employees, very bright people, who would like to get more involved on a volunteer basis within the schools."

Once they were persuaded that there was no hidden agenda, Johnson said, the principals of Pierce Elementary School, Wilson Middle School and Kennedy High School happily accepted the offer. What started as a pilot program in 1989 with those three Cedar Rapids schools was expanded to include all of the public schools in the district. It eventually became a nationwide Rockwell Collins program, one of many ways employees are involved in their local communities.

› Concern for employees' families led to establishment of the Child Development Center, the first corporate child care center in Cedar Rapids.

› Recreation facilities established in Cedar Rapids were used by thousands of employees and their families.

> Cedar Rapids Area Chamber of Commerce board chairman Jack Evans (center) presented Community Recognition Award certificates in 1985 to Rockwell executives (from left) James Churchill, Donald R. Beall, Jack Cosgrove and Arthur J. Rhodes.

The relationship between the Cedar Rapids community and the company was not always close, despite the Collins role as a major employer. The Cedar Rapids Area Chamber of Commerce did present its "Community Recognition Award" to Rockwell International in October 1985, recognizing all three Cedar Rapids-based Rockwell businesses: the Avionics Group, Collins Defense Communications and the Graphics Systems Division's plant.

Johnson recalled that city government, however, "kind of always took Collins and Rockwell for granted. Art Collins and the leadership of Collins really didn't pay much attention to the city; they were focused inward. There weren't many people in the Collins Radio organization who were very involved in the community."

He mentioned the issue to Lee Clancey, who later became the first woman mayor of Cedar Rapids. Churchill had expressed some dismay that city officials paid little attention to its largest employer. Clancey relayed the concerns to Mayor Don Canney, who quickly arranged a visit. Johnson said the influence of Rockwell International also helped.

"Rockwell had a little different corporate philosophy and wanted to become involved in the community," he noted. "As time went along, we got a lot more involved."

Coralville innovation

The GPS technology that had helped Chuck Hall and his crew make their historic flight across the Atlantic to Paris also led Collins to open a new manufacturing facility in Coralville. Designed to manufacture GPS end-user equipment and later expanded to include an automated warehouse and a microelectronics production area, the facility also became the proving ground for a cooperative new approach to getting the work done efficiently.

Soon after receiving the Air Force contract for Navstar user equipment, the Avionics Group had added 200 engineering and technical positions. An additional 600 would

> The Coralville manufacturing facilities, where CGAD and the IBEW established a unique cooperative working agreement, were designed with extensive employee input.

be added in production over the next several years to fulfill that contract and subsequent orders. The engineering offices and laboratories remained in Cedar Rapids, but officials looked at several sites, in Iowa and elsewhere, before selecting the Coralville site. Commenting on the decision in 1985, Jim Churchill said, "The corridor already possesses the infrastructure for high technology industrial development — interstate highway and air transportation, quality work force, excellent educational institutions, progressive cities and good utilities."

He also noted changes in the business climate as factors in the decision, including the availability of job training programs, and developers' willingness to commit capital. He commended the International Brotherhood of Electrical Workers (IBEW) for an agreement that would help keep costs down. Ground was broken August 21, 1985.

Ron Schleder, the first manager of the Coralville plant, said the High Performance Work Systems "team concept" that was implemented in Coralville called for high levels of employee participation in decision-making. It required development of a special agreement between corporate leadership and the IBEW to define only two bargaining-unit labor classifications: Assembly Operator and Test Equipment Technician. In contrast, agreements in force in Cedar Rapids included dozens of classifications.

Employees at every level were involved early on in Coralville, even helping design the layout of the assembly floor. When Schleder's team of hand-picked managers first saw the interior of the 100,000-square-foot facility, it was essentially empty. He credited them — and the partnership with IBEW Local 1634 — with getting the plant off to a quick, successful start.

A year or so after the Coralville plant opened, Schleder recalled, Product Support Teams were organized, and they soon became critical to successful operations. The teams included industrial and test engineers in addition to managers, assembly operators and test personnel in one of the organization's first moves to have engineering and manufacturing work more closely together.

Plants in the "Top 10"

The Coralville operation was different in more than just its level of automation and a new approach to cooperation between labor and management. It was altogether new, for example, compared to the Main Plant on 35th Street in Cedar Rapids. As it was being filled with new-fangled equipment and before the staff had grown much from a handful of managers to what would become several hundred employees, people sometimes brought their children to play in the big open spaces. The agreement between management and the IBEW led to unprecedented cooperation, sometimes to the consternation of their bargaining-unit counterparts in Cedar Rapids. Some Coralville employees describe themselves as "Coralvillains" and "a little weird" because they willingly train non-union workers at times when it is necessary to achieving company goals. Despite the differences, both ways of doing business seemed to work well. The Main Plant and Coralville operation were both recognized as "Top 10 Plants" by *Industry Week* in 1996.

Kent Statler, who managed the Coralville plant from 1996 to 1998, said the honor was awarded based on three factors that stood out: the High Performance Work Teams, the unusual labor agreement and its "living contract," and the people — especially the people. "It's your people that differentiate you," an *Industry Week* editor told him, "their engagement in the business, their knowledge of the business, their desire for the business to succeed." ▪

› Celebrating the "Top 10 Plants" honors were Cedar Rapids Mayor Lee Clancey, Jack Cosgrove, *Industry Week* Senior Editor Bill Miller and Herm Reininga.

"This new organization was very successful as each PST leader had all the resources working directly for him to accomplish all aspects of manufacturing and producing his assigned product line," Schleder recalled. "This proved to be one of the single best decisions we ever made at Coralville."

Kent Statler, who served as plant manager from 1996 to 1998, noted that the "living" nature of the company/union partnership agreement was another key to success: "Instead of having our contract in a sealed binder and locked in a safe, we had what was called the living contract, where as issues came up that impacted the business, the company and the union could get together and work on those. There would actually be revisions made out of cycle…to meet both parties' interests but at the end of the day to improve the performance of the business."

The level of automation introduced in the Coralville facility also was seen as ground-breaking. Herm Reininga, Operations V.P., noted that the introduction of automated processes in the defense and avionics industry had not been perceived as a good or even necessary thing. Providing avionics for companies that built just a few hundred aircraft a year for their airline customers, the Collins businesses and their competitors had been content to produce at low volume in traditional, non-automated ways. Reininga believed there were significant cost-savings to be had through automating even low-volume processes.

Reininga earlier had overseen the introduction of automation in Collins Defense Communications. By the mid-1980s, commercial production also was more automated. Automation reached its highest point later in the Coralville plant's automated storage and retrieval system (AS/RS) warehouse and in circuit board operations. Robotic equipment was introduced to place as many as 16 components per second and up to 60,000 per hour. The auto-insertion system could adapt quickly from high-volume orders to low-volume orders and back. Some boards were used in the same plant in production of hand-held GPS units and GPS products for precision-guided JDAM munitions and other surface and airborne uses. Others were shipped to various other Rockwell Collins plants.

> Herm Reininga was in charge of increasingly automated manufacturing operations.

Decentralizing manufacturing, rethinking procurement

In addition to the Coralville facility, other plants were opened or expanded in Iowa over a period of 10 years. In contrast with those in Cedar Rapids and Coralville, the other Iowa plants were non-union facilities from the start. The Collins Air Transport Division dedicated a 23,000-square-foot plant in Manchester in November 1986. The division's Decorah plant, which had opened in March 1978, was expanded from 20,000 to 35,000 square feet in late 1987 to increase production capacity. The expansion also allowed the company to bring all of the Decorah operations, some of which had been in leased space in downtown Decorah, under one roof. The division opened a new facility in Bellevue in June 1988.

Churchill said decentralizing the manufacturing and subassembly operations gave management "some leverage" in union negotiations, and it also served to enlarge the pool of potential employees. Reininga explained that part of the rationale was to limit the size of the Cedar Rapids workforce, which had numbered as high as 13,000 but in down times experienced layoffs also numbering in the thousands. Reininga said it was a burden on the community and costly in other ways, as well. A decision was made to cap employment in Cedar Rapids at about 8,000, in part to insulate the community somewhat from massive ups and downs. Reducing labor costs was also an objective. While most of the highly technical work and highly trained people remained in Cedar Rapids, some assembly work was moved to other Iowa communities and to operations in El Paso, Newport Beach, Dallas and Salt Lake City.

Salt Lake City also was the site of the company's first automated storage and retrieval system (AS/RS), which was later moved to Coralville. Reininga, who oversaw global production and material operations from 1982 until he retired in 2003, credited development of the AS/RS facility with greatly improved accuracy and savings in the shipping of parts for products to be assembled in plants at Bellevue, Decorah and Manchester in Iowa and others around the country. Previously, accuracy reached 92 to 93 percent. With automation, that reached 99 percent.

> Manufacturing facilities were expanded in Decorah, Bellevue (below) and other locations in Iowa, Texas, California and Utah.

The Chengdu F-7M tactical fighter manufactured in the People's Republic of China featured advanced navigation avionics developed by Collins personnel.

As production costs were reduced, the Collins businesses were able to bid more competitively and win additional GPS production contracts and others. Rethinking how to do business with the government — and with suppliers — would lead to additional cost savings and more profitable business wins.

Cosgrove recalled that the government procurement process in the 1980s often was influenced by "the program du jour" as the services responded to what was supposed to be the latest and greatest new manufacturing process: "We'd get a letter from the Army, Navy, Marines and Air Force telling us to submit our plans for Total Quality Management, for Quality Circles, for all that stuff. We kept saying, 'Why don't you just tell us what your requirement is and let us design something for that requirement rather than coming up with an RFP that's this big and you've got a specification down to what kind of resistors to use and capacitors to use and semiconductors to use, and let us take responsibility for that. We hammered at that for a long time."

The hammering would pay off over the next several years.

End of an era

As the 1980s were drawing to a close, the Collins units were increasingly looking abroad for new business. They teamed up to provide advanced navigation avionics for the F-7M tactical fighter manufactured for export by the People's Republic of China. The French government chose the Collins ARC-186 VHF transceiver for a new fleet of helicopters. A new licensing agreement meant some of that production would be done at the company's facility in Toulouse, France.

Other developments abroad — in the Union of Soviet Socialist Republics, in particular — would have a different sort of impact. The U.S.S.R. was beginning to come apart. With the longtime Soviet adversary no longer so menacing, the industry could not count on the U.S. continuing its same, high levels of defense spending. That particular boom was getting ready to go bust.

> The project team that developed the DARPA miniature GPS receiver met a challenge in miniaturization that made possible increasingly smaller, more sophisticated and more portable GPS products. Collins' GPS expertise contributed greatly to the company's success in the 1990s and later.

Back in the United States, other trends had their effects on the business. Among them were increasing environmental concerns. In July 1989, Rockwell International closed its electronics manufacturing facility in Lisbon, Maine, which had been in operation since 1972. The corporation cited the costs of complying with changing environmental requirements as a major factor in the closing, and it accepted ongoing responsibility for mitigating groundwater contamination related to the facility's operations.

Jim Churchill was preparing to retire, which he did at the end of 1990. He reflected on recent record sales and also looked ahead:

"We know there will be a shift in the commercial side, which is part of the natural cycle. But barring a deep national recession, we expect flat to modest growth in the commercial transport markets over the next five years. We expect the commuter market to continue to grow. In addition to more new aircraft, our average content per plane is also increasing."

He said he also looked for Collins' government business to continue growing in the face of overall flatness in defense spending, with GPS and JTIDS data link programs leading the way.

Later he would say he was largely pleased with his tenure as president of the Avionics Group, but would have liked to have seen significantly more success internationally. He noted that it was difficult to overcome foreign competition that was subsidized by foreign countries that wanted to maintain their own content — on the Airbus platform, for example.

"Our penetration in obvious areas such as communication was all right, but we weren't getting the subsystem penetration that I'd hoped we would get," he said. "It turned out to be better a little later."

Indeed, in time it would become much better. ▪

FD
ADF
DATA
2ND CRS
2000
ALT
AP
LNAV
20
10
200
100
2000
900
800
29.92IN
240
220
200
180
B 210
DG2
N
3
33
30
6
FMS 341°
10NM

Years of transition

Transition is a common theme throughout the history of Collins Radio Company, the Collins businesses as part of Rockwell International and, later, Rockwell Collins. Over time, science and technology seemed to change ever more quickly. Hardware-oriented business morphed into software-dominated, integrated systems. Electromechanical dials became CRTs and, later, liquid crystal displays (LCDs). Mechanical flight controls became digital, fly-by-wire marvels. Analog became digital. Voice communications evolved into data and video networks. And a leader in aviation electronics began to pursue more down-to-earth opportunities.

Many of the most notable changes in the Collins businesses themselves happened in the space of just a few years.

‹ Liquid crystal displays (LCDs) consumed less power and took up less space than those using cathode ray tubes and became critical to success in the competition for avionics business.

› Satellite-based systems enabled improved communications for commercial airlines as well as armed forces, and Collins SATCOM terminals became big business.

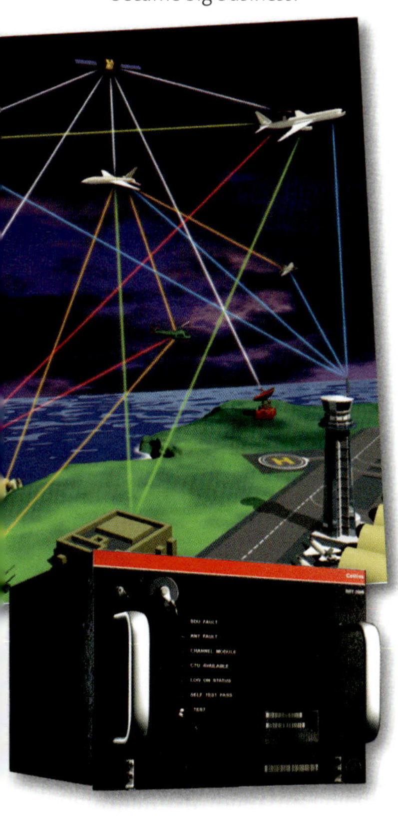

The Collins commercial avionics business set three significant records in 1989, with $1.1 billion in orders, $984 million in sales and a year-end backlog of $1 billion. It held a leading position in the commercial air transport market, with a 40 percent share of its addressed market, and continued to be a major supplier to Boeing, McDonnell Douglas, Airbus, British Aerospace and Fokker. Among the orders received were several for the new Satellite Communication (SATCOM) system for worldwide communication by transoceanic aircraft and others on "long-haul" routes. Orders were in place from United, Northwest, Cathay Pacific and Japan Airlines.

Those records had helped the corporation's Electronics Division become the largest segment of Rockwell International's business, outpacing Aerospace, Automotive and Graphics. As the 1989 annual report indicated, the growth had been coming increasingly in the commercial and international markets.

› As domestic airline expenditures declined in the early 1990s, maintenance contracts with major airlines were an important source of revenue. In one such agreement, CCA maintained avionics for American Airlines' fleet of Boeing, Airbus and Fokker 100 aircraft (above and cockpit at right).

Indeed, it noted, "The proportion of our business with the U.S. Department of Defense (DOD) is declining." Performance of the Defense Electronics segment had fallen short of expectations for the year. Projections were that there would be zero or negative real growth in U.S. defense budgets for the next several years. Competition was becoming more intense as the industry restructured, and some reorganization was about to hit home in Cedar Rapids, as well.

Rockwell's chairman and CEO, Don Beall, told employees at the Rockwell People Forum in 1990 that the company was "well positioned" with a long-term diversification plan that would "prove invaluable as we face declining defense budgets, increased competition and a globalization of many industries." He talked about teamwork and how the business needed to become increasingly market-focused and customer-driven. Those would all become increasingly important concepts to the Collins businesses.

John Girotto, who had been vice president and general manager of the Air Transport Division, was named to succeed Jim Churchill as president of the Commercial Avionics Group, which became known as Collins Commercial Avionics (CCA). He took the post in 1990 as U.S. airlines began a four-year slide in which they would lose nearly $13 billion. Girotto recalled that a new drive was on to increase shareholder value despite shrinking defense budgets. The corporation's stock price floated in the $20s; the goal was $100.

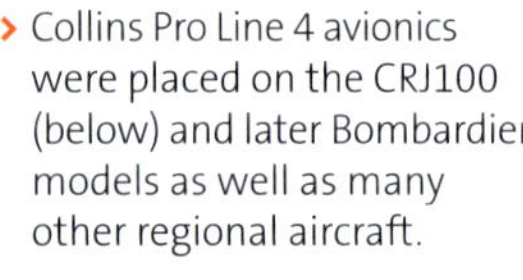
› Collins Pro Line 4 avionics were placed on the CRJ100 (below) and later Bombardier models as well as many other regional aircraft.

Lufthansa photo.

Looking for new markets, opportunities

Efforts to increase value led to looking at emerging markets both at home and abroad, most notably for the Collins businesses in growing markets for general aviation, business and regional transport aircraft, new opportunities in Russia and China, and in finding new uses for GPS and other technologies in land transportation.

Girotto said CCA took special note of new developments in the burgeoning regional aircraft market. In 1990, CCA held a 35 percent share of the general aviation avionics market, turbine aircraft operated largely by regional airlines and corporations.

› The CCA General Aviation Division was a leading avionics supplier for regional aircraft like the Saab 2000, which featured large LCDs on the flight deck (above).

"We were starting to see the effects of commuter flying," he said, and saw an opportunity to make a mark as a potential systems provider. Taking lessons learned in development of the L-1011, and armed with resources and experience from its successes in placing systems on the Boeing 757 and 767, CCA's General Aviation Division targeted Bombardier's regional jet program and other regional and business aircraft. The same division had developed the advanced avionics system for the Beech Starship several years earlier, leading to the increasingly popular Pro Line 4 avionics suite and the Collins division's emergence as the leading supplier of avionics to the business and regional markets.

Bombardier's Regional Aircraft Division selected the Pro Line 4 suite for its Canadair CRJ100 as well as later models. Pro Line 4 also was placed on the Beechjet, the Beech TTTS Jayhawk, the Saab 2000, Embraer's EMB-145 and its CBA-123. The regional CBA-123 was especially noteworthy as the first 19-passenger airplane to be built with an Engine Indication and Crew Alerting System (EICAS).

Pro Line 4 capabilities were enhanced with each subsequent program. The Beechjet version retained the basic Starship avionics but the classic "T" display was changed to a single primary flight display. An enhanced flight management display was added for the Canadair regional jet, and larger displays were added for the Saab 2000. Bob Hirvela, vice president and general manager of the General Aviation Division, pointed out in a 1990 *Rockwell News* article that the Beech T-1A Jayhawk military trainer installation was a joint effort with the Collins Avionics and Communications Division. Air Force pilots would be trained in the military version of the Beechjet to prepare them to fly KC-135 tanker aircraft.

"The government people are doing the engineering to adapt the Beechjet shipset to meet military requirements," he said. "It's working out very well. It's a good example of the cooperation between the government and commercial divisions." Such arrangements would later become much more commonplace.

In October 1990, the French aircraft manufacturer Dassault Aviation chose Collins Pro Line 4 cockpit avionics equipment over Honeywell for its new Falcon 2000 business jet.

› The Collins General Aviation Division and CACD worked together to provide Pro Line 4 avionics for the Beech T-1A Jayhawk, the military version of the Beechjet.

Honeywell's flight controls and displays had been standard on the Dassault Falcon 900, so the Pro Line choice turned heads when it was announced at the National Business Aircraft Association's convention. It was the eighth Collins win in a string of nine programs for which it had competed with Honeywell. Dassault was a recognized leader in technology, which added more luster to the company's growing reputation in the market. The Pro Line 4 system included four side-by-side CRT displays to present primary flight and navigation information. The Collins division also provided autopilot, flight director and centralized avionics maintenance functions through its Integrated Avionics Processing System.

United Airlines recognized Collins Commercial Avionics for its exemplary product and technical support by selecting CCA for its 1991 Maintenance Operations Vendor Selection Award. It cited the special support the Air Transport Division provided in demonstrating new air traffic control and data link capabilities for aircraft navigation. The airline had turned to Collins personnel to find a reliable way for flight crews to communicate with air traffic controllers on Pacific routes, where HF voice radio was sometimes unusable. In two months, engineers were able to provide a data link system using satellite communications rather than voice radio. It was later adapted for use on other routes.

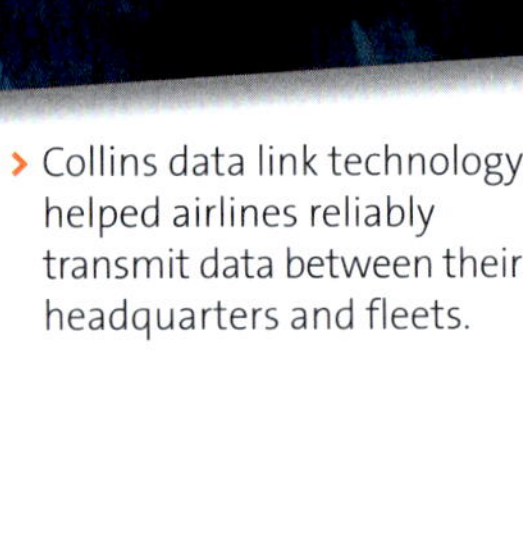

> Collins data link technology helped airlines reliably transmit data between their headquarters and fleets.

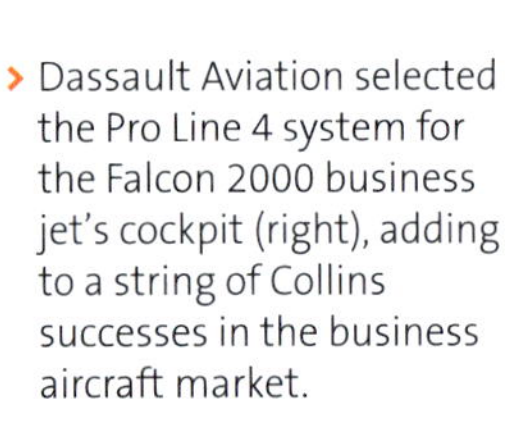

> Dassault Aviation selected the Pro Line 4 system for the Falcon 2000 business jet's cockpit (right), adding to a string of Collins successes in the business aircraft market.

The Air Transport Division also was continuing to advance data link and display technology with the Aircraft Communication Addressing and Reporting System (ACARS) used by airlines to transmit data between their headquarters and fleets. The Collins system, assembled primarily in Bellevue and tested in Cedar Rapids, consisted of two airborne units, a management unit and a control display unit. It also featured the most advanced display available in the industry: a liquid-crystal, touchscreen display. United Airlines equipped its 737s, 757s, 767s and 747-400s with the system, and Delta and other airlines also became major customers.

Rockwell International's annual report for 1991 told investors the company would have a "major presence" on the wide-body Boeing 777. In reality, Collins Commercial Avionics secured a relatively minor standard position on the 777 despite the selection of its Series 900 avionics products for the new aircraft by many individual airlines. CCA shipped some of that equipment well ahead of schedule in late 1993, including the traffic alert and collision avoidance system, Mode S transponder, weather radar, HF transceiver and satellite communications equipment. Also in development for the 777 were the autopilot/flight director system, a maintenance access terminal and optional side displays.

Historically, the division's relationship with Boeing had been strong and led to substantial positions on the 737, 757 and 767. Losing out to Honeywell on both the 777 and the "next generation" 737NG left CCA to look for victories in various other arenas.

› The Boeing 747-400 was one of many aircraft that used Collins ACARS equipment, which included touchscreen LCDs.

Boeing photo.

‹ With an agreement to provide a complete avionics suite for the Ilyushin Il-96M, Collins Commercial Avionics made history by becoming the first Western avionics manufacturer to do business in the Soviet Union.

Global marketplace changes

The Berlin Wall had come down in November 1989. While the Soviet Union was not formally dissolved until the end of 1991, the crumbling of the superpower was having its effects as a period of relative calm and security began to settle in. Competition for fewer defense dollars became more intense as the Cold War began to thaw, leading to mergers in an industry that had been flourishing for decades. At the same time, opportunities began to open in overseas markets for U.S. manufacturers.

The Avionics Group had already had some success in Western Europe. In 1990, it implemented an expanded "New Europe" marketing strategy that targeted Czechoslovakia, Romania and the Soviet Union.

› John Girotto was CCA president from 1990 to 1996.

Girotto later marveled at the striking changes in international relations. On one of many visits to Moscow, he recalled, it was "surreal," bouncing through the streets of Moscow in a Russian car with Don Beall, not long after the fall of the Berlin Wall.

"Here we are, sitting in Moscow," he recalled, "and we're about to start a business arrangement." Beall, Girotto and others were about to fly out to tour an Ilyushin aircraft facility. Ilyushin would soon become a customer for avionics. Girotto and others had also visited China a number of times over the years, trying to develop a relationship. Girotto noted some distinct differences between the business cultures in China and the soon to be former Soviet Union, both of which had been isolated from Western influences. Despite the "closing" of China to the West for several decades of the 20th Century, the Chinese had long been involved in international trade.

"The Chinese have been a part of commerce forever," Girotto said. "They were traders; they knew how to conduct business. When you went to Russia, they didn't know anything. They had no idea what to do and we were there trying to teach them. It was amazing, the contrast between those two."

The Russians, obviously, did know how to build airplanes, although they generally were heavy by Western standards and consumed much more fuel. The Russians were also willing to deal. By October 1991, with U.S. airline expenditures falling (and with declining earnings overall in CCA), Collins Commercial Avionics was picked to provide the complete avionics suite for the Ilyushin 96M, a wide-body aircraft that the manufacturer hoped to sell both inside the Soviet Union and in other countries. The Air Transport Division was selected to be the systems integrator and supplier of all the avionics equipment for the 375-passenger jetliner.

The historic Il-96M agreement involved a display system combining an Electronic Flight Instrument System (EFIS), Engine Indication and Crew Advisory System and Flight Management System; Flight Control System; and other avionics including a maintenance system, digital flight data recorder and satellite communications. Operating jointly with the Russian State Scientific Flight Research Institute (GosNIIAS), CCA employed more than 100 Russian software engineers to customize the Collins avionics software for the Il-96M. The first shipment of avionics for the Il-96M was made in July 1992.

With the Ilyushin agreement, Collins Commercial Avionics became the first Western avionics manufacturer to do business in the Soviet Union. Soon after, in 1993, it was

› Flight decks for the Ilyushin Il-96M (top) and Tupolev Tu-204 (bottom) featured avionics adapted from those used in the Boeing 747-400. The agreements were followed by others with Russian airlines to equip Boeing 737s.

selected to provide the avionics suite for yet another Russian aircraft: the Tupolev 204. The agreement with the British-Russian Aviation Corporation meant the Air Transport Division would supply a digital autopilot system; the EFIS, with six 8-inch CRTs for primary flight displays, navigation displays, and engine indication and crew alert displays; integrated flight controls, flight management systems and onboard maintenance functions; and TCAS and weather radar systems as standard equipment. Avionics for both the Il-96M and the 214-passenger Tu-204 were adapted from the suite developed for the Boeing 747-400.

Three GosNIIAS software engineers spent nearly six weeks in Cedar Rapids in the summer of 1992, getting familiar with the avionics equipment for the Il-96M. While the Russians were learning about American culture and technology, some Collins Air Transport Division employees spent months in Russia working to adapt avionics to the new market. The relationships established with the Ilyushin and Tupolev programs later paid off when Russian airlines invested in lighter, more advanced Boeing aircraft. “When we first started seeing the Russians buying the 737, everything that could be selected was Collins,” Girotto said.

Numerous missions to China, and establishment of a service center there, also began to bear fruit. Lacking the aircraft manufacturing experience of the Russians, the Chinese arranged to build MD-80 aircraft under a licensing arrangement with McDonnell Douglas and used all Collins buyer-furnished equipment.

Back in the United States, the Collins Air Transport Division reached another milestone in late March 1992 when the 5,000th WXR-700 weather radar system came off the assembly line, 12 years after its introduction. It was a number not often reached by avionics products because of the limited number of aircraft and rapid changes in technology. The 5,000th system was delivered to Northwest Airlines to be installed on an Airbus 320. Northwest was one of 145 airlines that had adopted the WXR-700, a system unique in its ability to operate on low power — 125 watts in X-band and 250 watts in C-band – when other radar systems typically required 65 kilowatts. It was also the first solid state, digital radar system for commercial aircraft.

Company transformation

Under Jim Churchill, the Avionics Group had included the Collins Government Avionics Division (CGAD) in addition to the commercially oriented General Aviation and Air Transport divisions. CGAD had developed two of the most important Rockwell International relationships with the Department of Defense, in the form of GPS user equipment and the JTIDS data links program.

After Churchill's retirement, CGAD was moved into Collins Defense Communications (CDC), which was reorganized in 1990 as the Collins Avionics and Communications Division (CACD) with Jack Cosgrove as president. The Dallas-based Communications Systems Division was merged into CACD, as well.

Girotto recalled that one purpose of the reorganization was to "keep all of the defense activities clustered" because of uncertainties about the future of defense budgets. Perhaps more important, the reorganization did away with part of a longstanding divide between the Collins businesses, one step in what would be a continuing, sometimes tumultuous process. There were no longer two, sometimes competing divisions targeting the Department of Defense from different perspectives.

The change in 1990 was later described as a turning point, a "pivotal" moment, by Robert "Kelly" Ortberg, who joined the company in 1987 and later served in several key management roles.

"Looking back at that, we had two different organizations, radically different organizations, addressing the same market space," he said. "We had two defense businesses addressing the U.S. DOD communications market" — one in CGAD and one in CDC, on opposite sides of C Avenue. "That move was when Collins created a strong defense, government systems business. That's what I call the birth of the modern-day Government Systems business."

At the same time, the move drew a more solid line between the company's government and commercial businesses, which had developed distinctly different cultures as separate entities inside Rockwell International. C Avenue in Cedar Rapids was a symbolic dividing line, with commercial business on the west and government on the east. Some described it as a "Chinese wall" and others "as wide as the Pacific Ocean."

"C Avenue was maybe 80 feet wide but it was 40 miles high," Cosgrove recalled. "The whole Collins culture was different in both camps. We were in the same city, right across the street facing each other, and our business cultures were completely different." That extended to manufacturing and quality control standards as well as how each unit related to its respective customers. Rigid government specifications and procurement procedures dictated to a great extent how the defense business operated in contrast to Commercial Avionics in the civilian aviation market. The defense side had a reputation for being more rigid, as well, compared to the relatively flexible commercial organization. The two had some manufacturing operations in common, but each division had its own financial and engineering organizations. They also reported to the corporation through distinctly different arms of Rockwell International.

› James L. Churchill was president of the Avionics Group from 1981 to 1990.

› Originally a program manager for the Multi-functional Information Distribution System (MIDS) program, Robert "Kelly" Ortberg became Rockwell Collins Commercial Systems executive vice president and COO in 2006.

Comm Central links service members and families once again

Collins people and technology have long been instrumental in helping military personnel communicate to complete critical missions. In 1990, they restarted their participation in the Military Affiliate Radio Station (MARS) program to help U.S. service members stay in touch on the home front. CACD and the Collins Amateur Radio Club teamed up to provide a 24/7 link between the United States and the Persian Gulf during Operation Desert Shield. Employees, including many volunteers, provided a free radio and telephone link between service members and home. While the Comm Central facilities had since been relocated from Building 121 to Building 120, they had also been used to help soldiers, sailors and airmen stay in touch with loved ones during the Vietnam War. ▪

› CACD and the Collins Amateur Radio Club teamed up in "Comm Central" to help soldiers, sailors and airmen call home from around the world.

With the realignment of the Collins businesses, Rockwell International had more clearly delineated their respective markets for years to come. CACD was synonymous with "government" and, more specifically, "military." The newly renamed Collins Commercial Avionics (CCA) was more definitively commercial, and freed from the onerous Pentagon accounting rules.

Exploring new markets

Faced with slowdowns in defense spending, Rockwell International was increasingly turning to commercial ventures. CCA, too, was looking beyond aviation as the commercial airline business slowed. Despite the aviation orientation of its large Air Transport and General Aviation divisions, the division competed for land-based GPS and electronics business in the trucking and railroad industries.

In March 1990, the company announced its entry into the long-haul trucking industry with the formation of the Rockwell Mobile Communication Satellite System organization. The new venture — one of many in various Rockwell International businesses — was aimed at capturing a portion of what was estimated to be a $2 billion long-haul trucking market for satellite communications applications. In May 1991, Collins Commercial Avionics and CRST, Inc., a leading trucking firm headquartered in Cedar Rapids, began working together to define a truck-based satellite communication system to collect and analyze data from trucks on the road.

Other promising markets had been identified earlier in shipping and railroads. In 1989, Burlington Northern ordered Collins radio link switch controls for 350 miles

› A Burlington Northern locomotive arrived in Cedar Rapids in May 1990 with Rockwell's Locomotive Analysis and Reporting System (LARS) aboard.

of track, and ordered data communications and performance monitoring systems for the railroad's Advanced Railroad Electronics System to be installed on 100 locomotives. A Burlington Northern Railroad locomotive fitted with Rockwell's Locomotive Analysis and Reporting System (LARS) rolled into Cedar Rapids for inspection in May 1990. The LARS installation monitored power, fuel and liquid levels, throttle position, temperatures and pressures. Other in-cab electronics technology deliveries were made to Norfolk Southern and Union Pacific.

Taking avionics and communications equipment developed for airlines and adapting it for use in railroads and other land transportation made sense, at least on the surface. The thinking was that "transportation is transportation," and there were similarities between railroads and airlines. Both required heavy investments of capital, for example, and consumed a great deal of fuel to operate. The differences, however, turned out to be significant, according to Girotto.

"When a freight train gets to its destination, it might sit there three more days before anyone touches it," he said. Timing isn't as critical as it is for commercial airlines, which have to move passengers quickly and are always seeking ways to increase fuel efficiency to lower costs. Traffic control and dispatching requirements are vastly different.

"We had a system that could get the train there on time, but we couldn't get the railroads to step up," Girotto said. CCA's forays into railroading and trucking — and a similar CACD venture in the agriculture industry — wouldn't last through the '90s.

Domestic airline expenditures declined substantially in 1992. The relative lack of lucrative new avionics contracts with major air transport manufacturers compared to previous years was such that a maintenance contract found its way into the Rockwell International annual report as a "highlight" of the year. In June of that year, Collins Commercial Avionics signed an agreement to provide maintenance service on avionics installed on American Airlines' Boeing 757 and 757, Airbus A300 and Fokker 100 fleets — some 250 aircraft. Such traditional fixed-price contracts that specified charges for repairing each avionics unit would give way in a few years to groundbreaking "pay for success" arrangements.

› CCA ventured into satellite-based communications for trucking fleets in the early 1990s.

› The Tripmaster system recorded data on engine performance and trip information for commercial vehicle operators.

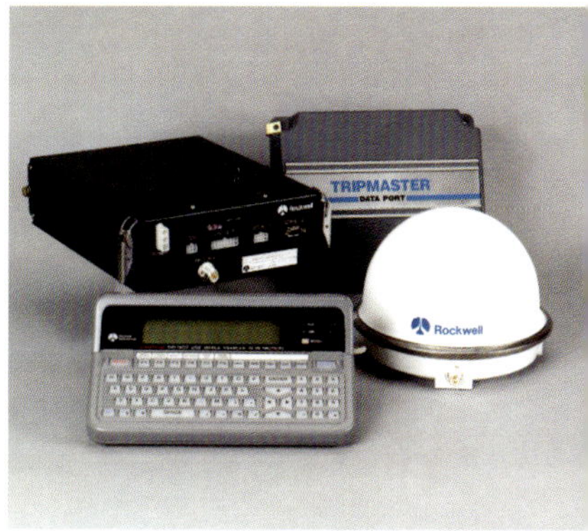

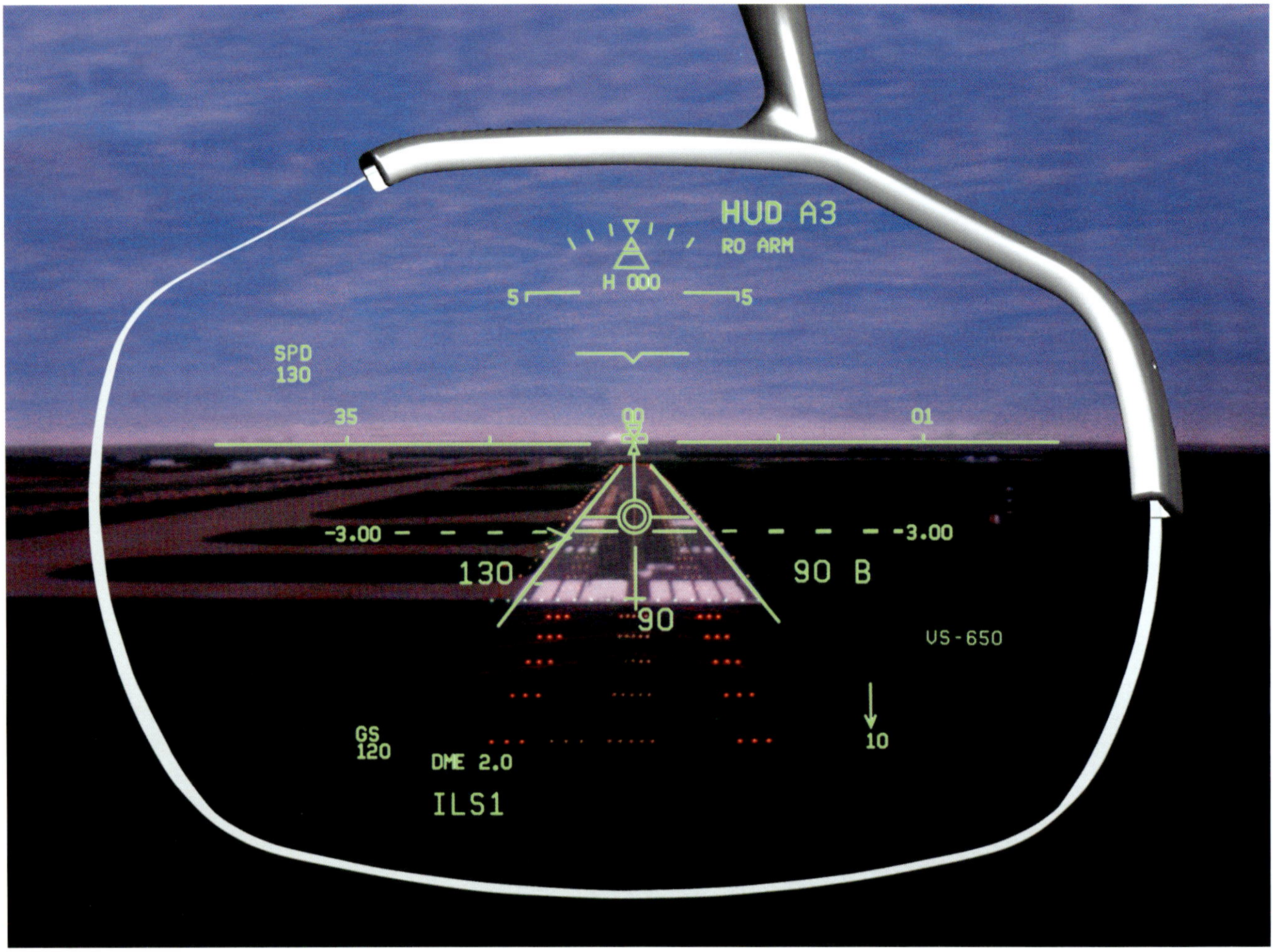

› Head-up displays put flight information directly in front of the pilot.

Early in 1993, CCA announced the formation of a joint venture company with Kaiser Aerospace and Electronics Corp. that would prove to be a key to enormous future breakthroughs in displays technology. Through the new company, the Air Transport Division and Kaiser would operate Flight Dynamics, which was acquired from Hughes Aircraft Co. Flight Dynamics, operated as a separate business unit based in Portland, Ore., manufactured patented Head-Up Guidance Systems™ for low-visibility approach and departure capability for air transport and regional aircraft.

Bob Tibor, vice president and general manager of the Air Transport Division, said at the time that the new technology provided "a natural extension of our all-weather landing and departure systems." Combining Flight Dynamics with Kaiser's expertise in military head-up display (HUD) capability and Air Transport's growing systems experience, the joint venture was expected to propel the team to a leadership role in HUDs and enhanced situational awareness in the air transport, regional and corporate aircraft markets. With a head-up display, a pilot can see a visual image of a runway projected on the aircraft's windshield when fog or other conditions would otherwise make landing impossible.

Soon after the announcement, Southwest Airlines placed a $45 million order for 236 head-up guidance systems from the joint venture company, with options for 63 more. CCA also had more than 380 orders for its SATCOM-906 satellite communications system. New customers for that system included Lufthansa, Iberia Airlines, Qantas and All Nippon Airways.

› Initially introduced for airline use, the SAT-906 satellite communications system also provided email, phone and fax connectivity to operators of business aircraft.

All Nippon, British Airways, Japan Airlines, Emirates Airlines, China Southern Airlines and Thai International Airways also had all placed orders for Series 900 advanced avionics equipment on their Boeing 777 aircraft.

In 1994, with the market for large, new commercial aircraft otherwise lethargic, CCA continued to focus on increasing its international business, as well as on commuter aircraft and longer-range business jets. The division became the first to certify a traffic alert and collision avoidance system (TCAS) in Russia, aboard an Aeroflot airliner. Worldwide, the division had orders for more than 3,500 TCAS II systems, which included all of Aeroflot's long-range passenger aircraft.

> Sales of the TCAS II system helped improve flight efficiency on oceanic routes and boosted CCA's international business.

The Air Transport Division had an opportunity in April 1994 to show the capacity of the TCAS system to improve flight efficiency, as well. With a United DC-10 and 747-100, it successfully demonstrated TCAS-assisted oceanic in-trail climb procedures to the FAA. It was the first use of the new air traffic control procedure to enhance flight efficiency on oceanic routes, where aircraft typically were assigned relatively fixed navigation tracks and had to maintain a minimum separation of 60 nautical miles to compensate for lack of surveillance radar on those routes. Using TCAS allowed a trailing aircraft to close to 15 miles at a higher altitude, thus enabling it to operate more efficiently and save potentially thousands of pounds of fuel on each flight. To use the maneuver, pilots had to establish the distance precisely to get air traffic control permission to use the maneuver. The successful demonstration was termed an "historic" achievement by FAA personnel.

Collins Flight Management Systems (FMS) were placed in service with Lufthansa CityLine, Comair and SkyWest regional airlines. The new FMS-6000 was selected as standard on Canadair's long-range business jet, the Challenger 604. Pro Line 4 was chosen by Israel's IAI for its new Astra SPX business jet — making it the choice at that point on 11 new aircraft, including planes from Beech, Canadair, Dassault, Learjet and Saab. CCA also introduced a new line of integrated, satellite-based precision navigation systems for business aircraft in 1994, using a newly developed 15-channel GPS receiver in combination with flight and navigation management capabilities. Collins AVSAT™ was introduced in October 1994 at the National Business Aircraft Association's trade show in New Orleans. Several different versions of the AVSAT were offered.

As the Rockwell International Collins businesses neared another turning point in their common history, they continued to expand their reach in 1995. The second largest airline in Europe, Lufthansa, had picked CCA as preferred avionics supplier for its fleet of Airbus 319 aircraft as well as future Airbus and Boeing 747-400s.

> The FMS-6000 was the company's first line of flight management computers that enabled pilots to easily calculate and see critical flight parameters at a glance.

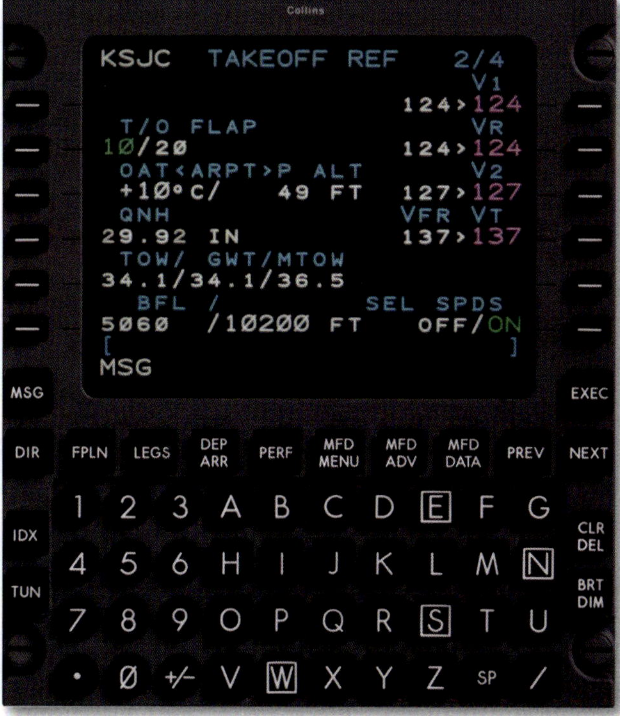

> The Pro Line 21 system was one of the first Collins systems to use liquid crystal displays in the cockpit.

CCA also partnered with Daimler-Benz Aerospace (DASA) to develop satellite-based flight guidance systems under the Collins Dasa Avionics Systems umbrella. The General Aviation Division had its first success with the new Pro Line 21 avionics system, one of the earliest Collins systems to use LCDs for cockpit displays, on Raytheon's new Premier I business jet. Pro Line 4 avionics were selected for the Dassault Falcon 50EX and the Bombardier CRJ-X.

Case for change

Jack Cosgrove recalled that CACD generally was feeling pretty good about itself in the early 1990s and had been winning about 40 percent of its bids — "which was not bad for defense business," he added. Still, a look at diminishing growth in defense budgets meant it was time to improve that rate and avoid costly mistakes in winning business that was not worth winning. The 1990 reorganization signaled a new era, in which everyone was going to learn: how to strategize, how to win business, and how to cut costs, even though not everyone saw the need in the beginning.

"We looked at the defense business at that time and the defense budgets were going down year to year, "Cosgrove said," and we sat there and looked at ourselves and, even though we were very successful financially, we said we've got to change the way we're doing business. In a down market like this, we've got to capture more market share or in effect take money out of somebody else's wallet, or we're going to go down with it."

Cosgrove said the learning began at the top, with a two- to three-day exercise on building teamwork. As people were trained and "empowered" to think for themselves at various levels throughout the division and even on the factory floor, some others felt threatened by their own loss of power and authority. Still, out-of-the-box thinking was encouraged and rewarded. Cost-cutting ideas resulted — seemingly simple things like paying parts suppliers based on shipping notices rather than waiting for them to send invoices — and were rewarded. "We saved a ton of money on that," Cosgrove said, and vendors were happy because they were paid sooner. Employees who came forth with ideas that would eliminate their own positions were assured of having a place found for them.

"We had different champions at all different levels of the organization, and it really took over and the people really started embracing this thing," Cosgrove added. Some remained skeptical, convinced that this new "Case for Change," as it was called, would soon go the way of previous improvement efforts such as the Total Quality Management and Quality Circles concepts.

"This is going to stick," Cosgrove insisted. Recalcitrants were told: The train is leaving the station. Get on board.

Lessons in teamwork were followed by additional training in how to respond to government RFPs and requests for quotes through a course called the Government Business Acquisition Process (GBAP). "We had to understand what the government wants, what the customer wants, and how best to respond to that," he said. The success rate climbed to more than 90 percent by 1996.

Cosgrove recalled that transition as a highlight of his career, demonstrating that the company was able to "change the culture of an engineering and manufacturing organization to embrace how to win and how to grow in a down business and a down market.

"We were worried that we would just follow that defense spending trend downward, so we started a process just after 1990 of cultural change. We went through a transition where we changed how we did business, got rid of all the silos that had been created, and started doing it based on teams and team performance."

Bringing some elements of the team concept together hinged, at least in part, on a sort of trick Herm Reininga played. Just as C Avenue separated division from division at times, there was a wall, every bit as real, that separated engineering from manufacturing. "One of the problems we had was getting engineers out into the factory," Reininga explained. "They were doing their thing, designing a product and then throwing it over the wall and operations would catch it. We'd go out and build it as best we could and have a lot of fun."

Reininga recalled that a visit from a carpet salesman, who was peddling carpeting for use in engineering spaces, sparked an idea. "I had stacks of reports on how you can't put carpeting in a manufacturing area," he said. Convinced it would work, though, he said, "We're going to do it anyway."

"In the engineering world, they have tile or old linoleum. Some places were just plain old cement polished up with a little wax. And I put carpet in the factory. You can't believe the number of engineers that went to the factory, not to help the factory out, but to see the carpet. And it broke down that barrier between engineering and manufacturing."

Girotto noted that a sort of "language barrier" also had developed as technology grew in sophistication.

"In the early '80s, most of our production folks were promoted out of the manufacturing ranks and our equipment was getting much more sophisticated,"

› Collins Pro Line 4 avionics systems contributed to the company's success in the regional jet market on aircraft like the Dassault Falcon 50EX.

Photo by Paul Bowen.

> The groundbreaking concept of Integrated Product Teams introduced in the early 1990s brought together engineering, manufacturing, finance and other personnel in support of individual product lines.

he explained. "The people that were being promoted didn't really have the technical background to interface with the engineering folks." Ultimately those difficulties led to some changes in hiring practices and more engineering expertise being brought into the manufacturing ranks through the early 1990s.

The transition from design to finished product was not always smooth, however. At times, it was far from efficient. Engineers naturally specified components to optimize each design, and plant personnel did their best to meet those specs. At one point, Reininga recalled, Operations had as many as 300,000 active part numbers — an enormous volume that represented a considerable degree of duplication. An engineer, with the best of intentions, would spec a component that would handle a 4 percent phase shift, for example, rather than 5 percent. While the same part would handle either spec from the supplier's perspective, two part numbers would end up applying to the same part, sometimes with two different prices. Partly as a result, in the late 1980s and early 1990s the company's cost of materials was rising 10 percent annually and it became increasingly important to find a way to better manage those expenses.

One avenue was to begin working more closely with suppliers through a Supplier Management Council, established in 1991 to understand and monitor changes in supplies of commodity parts, which become obsolete by the thousands every year. Another that same year was the creation of Commodity Teams, which included personnel from purchasing, component engineering, design engineering and quality control departments. Costs that had been rising 10 percent each year eventually started dropping at the same rate.

Reininga also pointed to the creation of Integrated Product Teams (IPTs) and Product Transition Lines (PTLs) in the early 1990s as groundbreaking steps that led to more efficiencies. IPTs were created to get engineering personnel more involved in manufacturing processes. As the concept evolved, it came to include manufacturing, engineering, finance personnel and others, all organized to support an individual product or product line. PTLs involved manufacturing personnel in the engineering process to help smooth the transition from design to product manufacturing on the factory floor. Throwing a design "over the wall" into manufacturing was a thing of the past.

JOINT TACTICAL
JTIDS
INFORMATION DISTRIBUTION SYSTEM

> Collins personnel were key players in the history of the JTIDS program. The radio-frequency network provides jam-resistant, secure voice and data communications for the U.S. military and its allies.

JTIDS and the power of networks

Developing the equipment and intricate software behind the Joint Tactical Information Distribution System (JTIDS) was an accomplishment that some compared to other landmarks in the history of Collins Radio Company.

"That was as innovative as Art Collins communicating by radio to the Pole, or the Apollo communications, and I think people don't appreciate that," said Kelly Ortberg, who became executive vice president and chief operating officer for Rockwell Collins Commercial Systems in 2006.

The radio-frequency information network provides constantly updated, jam-resistant voice and data communications between distant battle elements — Airborne Early Warning and Control System (AWACS) aircraft, fighters and bombers. It vastly improved pilots' awareness of their environment by extending their "vision" beyond the range of their onboard radar and their own field of view.

Making it work required jumping some significant technical hurdles, including the stringent spectrum-related requirements of the Federal Communications Commission.

Ortberg explained that "the radio frequency hopped 77,000 times a second, pseudo-randomly on large channel spacing, over 250 MHz of spectrum. This was the first time we had a radio hopping so fast across this frequency spectrum." The task was complicated further by the fact that the frequency range was also a navigation band.

"There were parts of that spectrum you couldn't have any energy in," Ortberg said. "It was a technological marvel how you could make a radio that did that and didn't violate any of the navigation channels and cause disruption in those channels. It really was complex."

He added that the JTIDS program proved to be "a pacesetter program" for Rockwell Collins: "What we do today is all around that network communication."

JTIDS terminals' primary function is to distribute tactical information in digital form. It operates over line-of-sight ranges up to 500 nautical miles with automatic relay extension beyond, locating and identifying friends and foes. The spread-spectrum and frequency hopping techniques make it resistant to jamming, and embedded data encryption makes it secure. A JTIDS terminal automatically broadcasts outgoing messages at predesignated, repeated intervals.

CGAD initially worked on the JTIDS program under a leader-follower arrangement as a subcontractor to Singer-Kearfott (which later became part of Plessey Electronics Systems, Marconi Electronic Systems, and then BAE Systems). A *Rockwell News* article in May 1990 noted that the division had won a portion of the work despite bidding higher than Plessey "because it has demonstrated impeccable quality and schedule performance." ▪

These transitional years also were memorable for the government's seizure of records from CGAD in 1989 and, two years later, a temporary suspension from bidding on or receiving government contracts. On November 21, 1989, Department of Defense agents from the Office of Inspector General, Defense Criminal Investigative Services, seized what was called "a massive amount" of records after sealing off part of Building 109. The investigation was related to quality and acceptance testing for the Air Force-sponsored Joint Tactical Information Distribution System (JTIDS), which CGAD had been working on since 1981. More than 100 prototype JTIDS terminals had been delivered by 1989. Ultimately, no criminal charges were filed.

An indictment in 1991, however, did charge two individuals and CACD with mail fraud and submission of false claims. The individuals were accused of altering time cards and fraudulently billing NASA for work on the Space Shuttle program. CACD

President Jack Cosgrove disputed the charges as unwarranted and said they involved procedural issues that were not criminal in nature. CACD was suspended from bidding on government programs in November 1991. Two months later, the suspension was lifted, leaving the company free to pursue government business again. A settlement announced that July required the company to pay the government $1.425 million and enhance its employee training regarding time-keeping procedures and ethics.

GPS production continued to contribute positively in 1989, despite gloomy expectations of smaller defense budgets. The Department of Defense exercised a $66.4 million option on the multi-year user equipment production contract that had been awarded in 1985. That brought the total awarded to date on that contract to $415 million.

Defense budgets in the early 1990s were flat or shrinking, and military customers were becoming more inclined to spend on modernizing existing weapons systems, like the U.S. Air Force's F-111F, than to start expensive new programs. Among other contracts was a $160 million award to update the avionics for the Royal Australian Air Force's F/RF-111C aircraft. The Department of Defense also awarded a $10 million contract for a smaller new GPS unit designed for military and international airborne markets, with options for an additional $90 million in production. Teaming with Canadian Marconi Company, Rockwell also received another $11.6 million contract to develop the Air Force's Military Microwave Landing System Avionics program.

› The Bell 406 combat scout helicopter was used to test the Collins Tactical Weapons Management System. The system was dubbed "Quick Draw" by pilots who tested it because it allowed them to rapidly select, arm and fire weapons.

In April 1990, CGAD introduced a new weapons management system to allow helicopter pilots to fire rockets, missiles or guns without removing their hands from the flight controls. The Collins Tactical Weapons Management System was a company-funded development program that combined off-the-shelf hardware with weapons software. It also controlled communications and navigation functions. Pilots who flight tested the system on the Bell 406 combat scout helicopter were so impressed with its capabilities they nicknamed it "Quick Draw."

› The C-160 cargo plane (left) and the F-15 (top) and F-22 fighters all used Collins avionics.

The initial JTIDS production contract, worth $59.1 million, was awarded in 1991. That followed a 1990 contract for low-rate production of JTIDS terminals for certain Air Force F-15s. (The bulky $2 million units were too large for most of the F-15 fleet.) Later, the company would provide smaller Multifunctional Information Distribution System (MIDS) terminals for additional F-15s and numerous other platforms.

The Collins government division continued to increase its share of the military market in 1992 with a $77.2 million contract for communication and navigation avionics for the U.S. Air Force F-22 Advanced Technology Fighter, and another $43.3 million in VLF business. That included nearly $36 million for continued preproduction engineering and logistics for the High Power Transmit Sets (HPTS) program. It also included development of an installation design to put HPTS on the E-4B national emergency Air Force command post aircraft. At the same time, the Advanced VLF Receiver (AVR) program was converted from development specifically for B-2 bomber avionics to supporting multiple programs, including the E-6A and E-4B.

Always known for top-quality, durable radio products, the Collins business grew those sales still further in 1994 with an $11 million contract to provide ARC-220 HF radios for U.S. Army helicopters, and another DOD contract for $44 million for ARC-210 Army, Navy and Marine radios. The U.S. Navy's airborne VLF communications program also continued to grow for CACD, with a $29.6 million HPTS order. The division also entered the ground VLF market with an additional $11 million order for the Navy's solid-state power amplifier program. Other key new business for CACD in 1994 included a $21 million contract to upgrade avionics on German C-160 cargo aircraft, $16 million for avionics upgrades on Canadian C-130s, and a $17 million contract for transceivers on Gulfstream IV and V aircraft. The U.S. Navy subsequently ordered $26.5 million more in High Power Transmit Sets.

The mainstay radio business continued to do its part, bringing 1995 follow-on production orders of $59 million from the Department of Defense for the ARC-210. The Air Force ordered 277 GRC-171 UHF radios for the sum of $23 million.

› The ARC-210 (above) and ARC-220 HF (below) radio system.

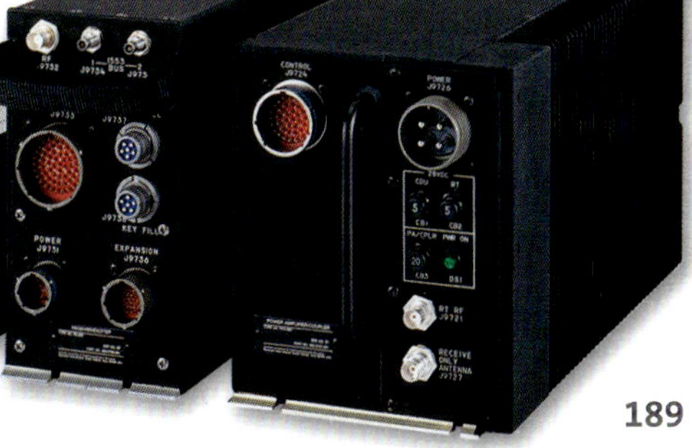

› The ARC-210 radio was chosen for the U.S. Navy F/A-18 Hornet.

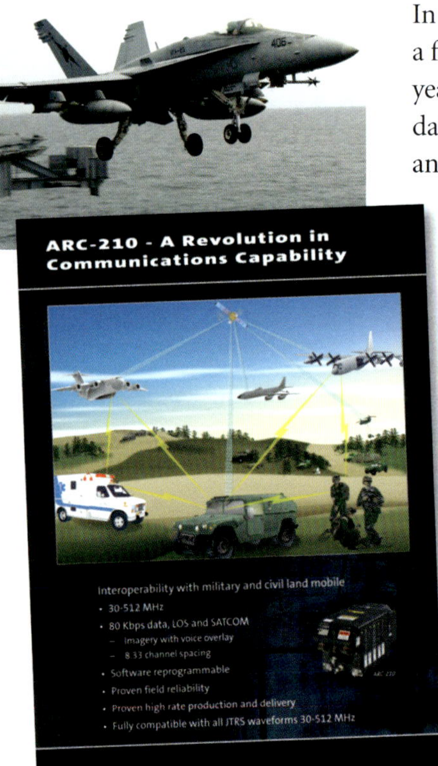

› The military adopted the ARC-210 radio as the standard for voice and data communications, helping make it one of the most widely used Collins products.

Changing procurement practices

In 1990, the U.S. Navy selected the Collins ARC-210 radio, which operated over a frequency range of 30 to 512 MHz, as a standard for multiple platforms. In ensuing years, it became the military standard for airborne, two-way, multi-mode voice and data communications, featuring embedded UHF and VHF anti-jam waveforms and other data link and secure communications features. At one key point, however, CACD faced the prospect of having to compete for follow-on production to reduce prices. In response to prodding from various interested parties, including Rockwell International and Collins management, the Department of Defense had launched an initiative to reform its procurement process. CACD was invited to participate as a partner in that reform.

The Pentagon had traditionally bought what it needed through a distinctively non-commercial process, specifying what and how a radio or other piece of equipment was to be made, down to precisely which parts would be used and how they would be labeled. Procurement, development, manufacturing and support processes were all steeped in bureaucracy. Bob Chiusano, who was named CACD vice president and general manager in 1996, said the system needed to be improved by making it "less elongated and more efficient."

CACD had been working with military customers to find a way to eliminate some of those layers of bureaucracy and the detailed specifications that previously dictated each step of design, production and support. Pentagon decisions to eliminate certain military specifications came along at a timely point. The government was leaning toward "buying commercial," particularly for what were basically "commodity" purchases like the ubiquitous radio. Government buyers, in those cases, seemed satisfied just to get an end product that did what it was supposed to do at a reasonable cost. The resulting agreement with the Navy enabled the division to reduce production costs and keep the business, including service and warranty work.

The division was awarded the next production contract in 1994 and eventually delivered more than 20,000 ARC-210 radios and generated more than a billion dollars in sales around the world. Chiusano summed it up: "The price of the ARC-210 dropped, our order book filled up because they could buy more, and the profitability on the unit went up because we were able to capitalize on economies of scale." The five-year production contract called for 10 percent annual reductions in price. At the same time, the expectation was that the mean time between failures (MTBF) would climb from 1,000 hours to 5,000 hours. The MTBF goal was met the first year.

› The ARC-210 "family" of equipment included several variants of the receiver-transmitter.

Chiusano said reform soon became "a way of life," a new, accepted way of doing government business in a more commercial fashion. "We were put on the map as a company that helped the government develop a reformed procurement process. It wasn't foreign anymore, we got used to doing it."

The Pentagon's increasingly flexible procurement process was reflected in GPS procurement when the Army began to respond to repeated suggestions that it give its contractors more responsibility for meeting product performance requirements and determining what specifications would be required. The government's willingness to consider buying essentially "off-the-shelf products" that were developed with traditionally commercial thinking behind them was demonstrated again in the pursuit of the handheld GPS business, which helped move that revolution along.

"We crafted a program that was time-phased over the years that said we would control the configuration, that we would warrant the equipment, that we would service the equipment," Cosgrove recalled. "We would do all of that for the military and all they had to do was buy it, just like going down to Radio Shack and buying a GPS receiver. By doing that, we could control the configuration, then as new state-of-the-art technology came out we could insert that any time we wanted to in the factory over the multi-year procurement. And they bought that."

"That" was the Precision Lightweight GPS Receiver — the PLGR. Earlier GPS miniaturization efforts were put to work in a smaller box — one that could fit in a soldier's hand or pocket. The need for a small, lightweight GPS unit had become clear during the Persian Gulf War, when U.S. troops had to operate in a desert environment that wasn't well mapped and lacked the geographic reference points available in many earlier conflicts.

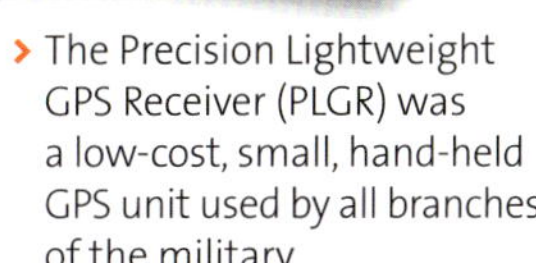

› The Precision Lightweight GPS Receiver (PLGR) was a low-cost, small, hand-held GPS unit used by all branches of the military.

CACD landed a $140 million contract in 1993 to produce up to 95,000 PLGR units. Pentagon officials who were overseeing the program didn't all seem to understand at first what had changed. They were preparing to review CACD's configuration management system and other program details but were told, "No, you didn't buy that. You're buying a commercial product."

› The Miniature Airborne GPS Receiver (MAGR).

"It didn't sit very well, but it stuck," Cosgrove said. "It changed the way government does business." He recalled that Gen. Otto Guenther made a speech later at the Coralville plant and told the assembled workers that he owed one of his three stars to the procurement process used in the PLGR agreement. The 20,000th PLGR unit was delivered 1994. That year, the Pentagon and international customers placed additional production orders totaling $47 million for approximately 27,000 units.

The government ordered $21 million more in PLGR handheld GPS units in 1995 and took delivery of the 50,000th PLGR. The Coralville plant eventually produced more than 250,000 PLGR units for the military and even more successor GPS products. It also produced another small "off the shelf" GPS unit in the 1990s, called the Miniature Airborne GPS Receiver (MAGR), for use in fighter aircraft and helicopters that weren't large enough to accommodate standard-sized airborne GPS receivers.

Both Collins Avionics and Communications Division and Collins Commercial Avionics participated, with four other Rockwell International business units, in developing still smaller GPS technology for OEM applications. The NavCore V unit — an "engine" that could be used in both military and civilian applications — was

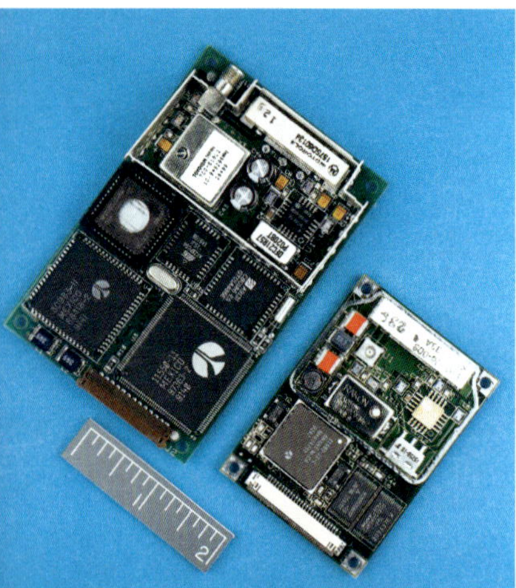

› NavCore V GPS receiver modules were small but powerful.

billed as "the world's smallest and most powerful 5-channel GPS receiver module" at just 2.5 by 4 inches. It was unveiled at a press conference in New York City in April 1991.

While the PLGR program may have played a role in the evolving government procurement process, it may have been more significant for how it changed the business itself. Big-box radios and avionics systems typically had been manufactured in small quantities for a total of a few hundred aircraft. Airborne GPS units were far too bulky to carry around. To win the PLGR business, everything had to be different.

Having to deal with the concept of ergonomics so users could easily operate a downsized GPS device was an entirely new experience. Not only did the technology have to be produced at a rate of hundreds per day rather than a few hundred per year, it had to be inexpensive — compared to a $20,000 radio, for example — and it had to fit in a soldier's pocket.

Integrated military systems

The company won an Army contract in May 1989 to provide ARN-149 low-frequency automatic direction finders (ADFs) as backup navigation systems in Apache helicopters. A relatively simple system, the ARN-149 was designed to home in on any radio transmitter operating in the 100 to 2200 kHz range. It was essentially a sale of an off-the-shelf, mature product that had found a small but profitable new market, and an illustration of the way the Collins Radio Company of old had operated. Arthur Collins' reputation as a technological genius lived on, as did an inclination to design and build a product and then find a market for it. (Not coincidentally, many of those products came to have useful lives measured in decades because of their rugged, quality construction.)

› The Army's Apache helicopters were equipped with Collins ARN-149 automatic direction finders for backup navigation.

With its selection as prime integrator for KC-135 avionics upgrades, the company stepped up the "food chain" from product manufacturer to systems integrator. The tanker aircraft refueled other aircraft in mid-flight.

As the next several years would show, by contrast, integrated systems developed by the Collins businesses — as opposed to individual products — would be a growing presence in both fixed-wing and rotary-wing markets.

A hybrid, off-the-shelf approach paid off in winning major business programs as the prime integrator for the Air Force's KC-135 avionics upgrades. Then 35 years old, the KC-135 tanker aircraft's mission was to refuel other aircraft. The first program, dubbed Pacer CRAG (for Compass, Radar and GPS) was awarded in October of 1995. It would be followed later with the KC-135 Global Air Traffic Management (GATM) program. In addition to a complement of military avionics and GPS technology, the winning proposal used weather radar and flight management systems developed for commercial airlines, a prime example of the increasingly common "dual use" concept in which technology developed for one market is leveraged effectively in another.

The KC-135 program's greatest significance, however, was that it marked the Collins unit's arrival as a systems and subsystems integrator — a difficult but successful step up the "food chain" that helped overcome the company's reputation for being just a reliable product manufacturer. Rockwell International very nearly took a pass on the opportunity.

Bob Chiusano said decision-makers had been reluctant to get heavily involved in the aircraft systems integration business.

"We had a group called Cockpit Flight Management Systems back then, but it was floundering," he recalled. "We weren't making any money." Ultimately, however, executives recognized that "moving up the food chain from a product supplier to a subsystems and systems supplier was something that was important to our long-term strategic perspective."

Lacking a track record in systems integration, however, the business unit was a long-shot to provide a fully integrated cockpit flight management solution. The results achieved by an aggressive, focused business pursuit team took some industry players by surprise. The team produced a technical proposal that drew all "blues" — the highest rating in the Air Force's color-coded evaluation and rating system — and "knocked the cover off the ball," as Chiusano said. "We gave them a cost schedule, risk profile and set of performance criteria that were second to none."

ANGLE OF ATTACK
T/O
PER
53.9
TCN
01.2
TCN
DTK
169
MSG
FMS
354/0
←TCN
TAS
ALT
1000 FT
100 FT
IN HG
2992
PER
53.9
TCN
01.2
279
DTK
169
MSG
AMA
GCK
TCN
160
WIND
354/0
PER
FMS
←TCN
TAS
ABOVE
RNG 10
TA ONLY
+04
+20
+27

> Two Collins divisions provided advanced cockpit avionics (facing page) for the KC-135 upgrade program, which modernized the fleet of tanker aircraft and set a new standard for procurement.

CACD was the prime contractor. Collins Commercial Avionics was a subcontractor, as was Rockwell's North American Aircraft Modification Division. Bidding against Lockheed-Martin, Raytheon and Boeing, the team won the Pacer CRAG program with avionics equipment from both sides of C Avenue. More than a third of the cockpit avionics for the KC-135 upgrade came from the Collins commercial business.

"It taught us how to be a systems integrator," said Greg Churchill, who was involved in landing the KC-135 program and later became executive vice president and chief operating officer of Rockwell Collins Government Systems. "It gave us the opportunity to be the prime integrator in a competition with Boeing, responsible for everything, from the initial design through the installation and the support. It gave us the credibility to be an integrator."

The Pacer CRAG contract was the first awarded under the Air Force's new "Lightning Bolt" procurement streaming initiative. The program was praised as having set a new standard in government/industry relations.

"The teamwork between the contract/government team on Pacer CRAG is probably the strongest I've ever been associated with in the 20 years I've been in acquisition," said Col. Francis Veldman, the KC-135 systems program director, who was quoted in an article in Collins' *Vision* magazine. The program was expected to cost more than $90 million less than it might have under older procurement processes.

The KC-135 program's success and the company's growing reputation as a systems integrator set the stage for a string of other successes. Collins avionics and display systems, for example, were later selected for numerous Communication, Navigation and Surveillance/Global Air Traffic Management (CNS/GATM) cockpit upgrades for C-130 aircraft. The C-130 is used primarily to move troops and equipment for armed forces around the globe. ▪

CHAPTER 14

Back to core business

Beginning in the 1970s and on into the final decade of the 20th century, Rockwell International had grown into a diverse and increasingly commercially oriented global enterprise, with smaller and smaller relative emphasis on defense business. In 1984, only 39 percent of its total sales were commercial and international. From 65 percent in fiscal year 1994, that grew to 72 percent of $12.9 billion in total sales for 1995.

The corporation's makeup changed even more dramatically in 1996, when it announced the signing of an agreement for Boeing to acquire the Rockwell Aerospace and Defense businesses for $3.2 billion. Earlier in the year, the corporation had announced the sale of Rockwell Graphics Systems, with manufacturing and operations in Cedar Rapids.

‹ An initiative to examine enterprise manufacturing processes was undertaken in the 1990s.

› Below: An advanced flight deck "rig" used for testing of new open-architecture-based avionics and pilot interfaces and for demonstrating the new systems for customers.

Four key businesses remained: Avionics & Communications (A&C), Automation, Semiconductor Systems and Automotive. Within months, however, came the announcement that the automotive parts component would be spun off to become an independent company. Two years later, in June 1998, the semiconductor business went the same way. Both Rockwell International and A&C were operating under a new strategic direction of refocusing on core business.

For the time being, the businesses that traced their roots back to Collins Radio Company remained part of Rockwell International, but with a big difference: The organizational walls that separated them were coming down. In November 1996, the two divisions operating under the Collins Commercial Avionics banner — the Air Transport Division and the General Aviation Division — were consolidated with CACD and the Communications Systems Division based in Dallas. Jack Cosgrove became president of the new A&C business. John Girotto, who had started with Collins Radio in 1960, retired after serving as president of Collins Commercial Avionics since 1990.

The commercial and military businesses had been split some 21 years earlier, in large part because of customers' requirements and other differences in those markets. Separate infrastructures were developed to accommodate the distinct needs and demands of each. By the mid-1990s, however, a key factor in driving them apart was diminishing. Defense Secretary William Perry in 1994 had launched a massive review of military standards known as MILSpec Reform. The push for adoption of non-government standards was furthered with the 1995 passage of the National Technology Transfer and Advancement Act, which took effect in 1996 and required the use of private-sector standards where possible.

The timing was right to bring the businesses back together. The consolidation and reorganization also set the stage for their eventual spinoff as a standalone, independent business.

New senior leadership

A few weeks after the consolidated Avionics & Communications business was introduced in September 1996, it announced the new Rockwell Collins organizational structure and senior leadership team.

Jack Cosgrove, president, named Clay Jones as executive vice president "to provide leadership for all Avionics & Communications divisions as they pursue long-term global growth in their businesses."

Jones continued as acting head of the Air Transport Division. Executives named to key positions reporting to him were:

- Bob Chiusano, who was appointed V.P. and general manager, Collins Avionics & Communications Division.
- Bob Hirvela continued as V.P. and general manager of the General Aviation Division.
- Glenn Andrew was named V.P. and general manager of the Communication Systems Division.
- Dan Chadwick was named V.P. and general manager for New Business Initiatives.
- Bob Marovich was appointed V.P. and general manager for Collins Subsidiary & Service Businesses.
- John Cohn was appointed V.P., Global Strategy Development.
- Larry Erickson was named vice president and controller for A&C Finance.
- Herm Reininga was named V.P. Operations, responsible for all A&C production operations.
- Bill Richter continued as V.P., Human Resources.

› Jack Cosgrove led Rockwell Collins A&C until his retirement in 1999.

Said Dick Johnson, "In essence, what we really did was form the old Collins Radio Company again."

Cosgrove later recalled the merger as a personal career milestone: "I was there in 1975 when they broke apart. In 1996, they brought it all back together and I got to be there and be a part of it. To put it all back together the way it was in '75, it was just a thrill."

Working through the details of restructuring took time and effort. Post Merger Integration (PMI) teams were formed to identify best practices and systems that could be adopted throughout the enterprise to accommodate both commercial and military customers' needs. Working together had not always been done easily when the divisions were formally separated. In the new structure, it was encouraged and expected.

› Clay Jones was named executive vice president of Rockwell Collins A&C in 1996.

Clay Jones, who had joined Rockwell International in 1979 and served as the corporation's vice president of Aerospace Government Affairs and Marketing from 1988 to 1995, was installed as vice president and general manager of CCA's Air Transport Division in 1995. He retained that position temporarily after being named executive vice president of the merged business in 1996.

Early in the integration process, Jones identified some of the issues facing the new organization in an article in A&C's *Millennium* magazine. He said he was seeing "a great spirit of cooperation" and added, "The barriers, especially between the commercial and military side, are being broken down."

He added, "Any time you introduce change in any organization, you have to deal with people who are comfortable with the status quo. Introducing change can be challenging for us in A&C simply because our individual businesses are so successful. For that reason, it can be hard to convince people of the need to change." He said it would be important to clearly communicate the case for change.

Preparing to work together

The new company — made from others that had linked but divergent histories — needed a new business model, which would include a new vision roadmap and a "shared services" concept. The vision included two important words: Working together.

Bringing those words to life in just one critical area of the business would take an investment of more than $150 million. As separate business units for just over two decades, the Rockwell Collins businesses had sprouted numerous, diverse information systems. There were half a dozen order management systems in 1996. There were five manufacturing resource planning systems. Among still others, there were three finance systems: one for government business, one for commercial business, and one for operations.

John-Paul Besong, who headed the team of more than 100 individuals tasked with building a unified information system, would later say they were all "beautiful systems that worked well in splendid isolation."

› Jack Cosgrove (left) and Clay Jones discussed the state of the Rockwell Collins business in employee meetings in the recreation center.

Photo by David Perry.

> John-Paul Besong led the effort to select and integrate a new Enterprise Resource Planning system and later became senior vice president of Rockwell Collins e-Business.

"There were silos within silos," he recalled.

Besong pulled together internal experts from throughout the new organization to form the aptly named "Project Fusion" team. Begun in 1997, its mission was to develop an Enterprise Resource Planning (ERP) system to integrate the flow of business information into one set of processes, a common database, and a single set of information system training materials.

Project Fusion resulted in the selection of Systems Applications and Products (SAP) software for use throughout the organization. Following months of planning, coordination, rehearsals and training, the SAP system went "live" in the Business and Regional Systems business area in May 1999. The switch was thrown in Air Transport in mid-2000 and in Government Systems in October 2000.

Besong, who later became senior vice president of e-Business, said it took about six months for employees to fully adapt to the new system. "When you go live with a system like this, even longtime employees become new employees. It's almost like having people with 10, 20 years experience going back to Day One because now they're using a system that's very foreign to them. You can imagine how much chaos that unleashed."

The consolidation and updating of the information management system was estimated to have resulted in $50 million in reoccurring annual savings. More important, Besong noted, "It was a strategic necessity for the businesses to work together. You cannot put a price on that."

Bringing the businesses together was a massive, complicated effort by thousands of other employees, from executives and engineers to factory workers. The new leadership quickly established 25 PMI teams and numerous sub-teams to link and build on the efficiencies and strengths of the various pieces of CACD, Air Transport, General Aviation and CSD. Dan Chadwick, newly named to the post of vice president and general manager for new business initiatives, was given the additional task of coordinating the teams' efforts as A&C's PMI executive.

"Many of the people selected for PMI team efforts were involved in activities critical to the future of our business, and these folks were already carrying a heavy workload," he said early in 1997. "What we asked for was that 'extra effort' that characterizes winners; these people responded."

One outcome was increased interaction between engineering and manufacturing. That included enterprise-wide adoption of the Integrated Product Team (IPT) concept, which physically grouped all of the product-related functions — from engineering to scheduling to production and testing — to promote on-the-spot interaction and problem solving. Production workers were expected to become more accountable and more empowered, and they would get a better view of how their work contributed to the overall manufacturing process.

The consolidated businesses made many other significant changes to improve profitability. Removal of some of the barriers between businesses was accompanied by consolidation for efficiency of operations. Process centers that were operating separately in Cedar Rapids, Manchester and Coralville — all making circuit boards for different units — were consolidated in the Coralville plant, which had initially opened primarily for GPS production. It was one of the "first major steps" in bringing separate businesses together and in moving further toward a dual-use philosophy and facility, according to Kent Statler, who was the Coralville plant manager at the time.

"The machine doesn't care whether it's building a commercial board or a military board," he explained. "It will place the part, it will reflow the solder consistently. We used the same workmanship manuals. We were able to get tremendous leverage by bringing commercial and military together across a common process center."

Another result of the integration efforts was the establishment of the Advanced Technology Center to identify and develop technologies that could be used to advance the business and meet customers' needs. It would fund research into promising ideas developed by Collins engineers. Also recognizing that valuable technologies often originate elsewhere, it would develop partnerships with a variety of outside parties, such as the Marine Corps Warfighting Laboratory, the Air Force and Navy research laboratories, and DARPA.

› Dan Chadwick, then vice president and general manager for new business initiatives, was tasked with coordinating the work of Post Merger Integration teams when the Collins units were brought back together in 1996.

Turning points with Boeing

Consolidating the businesses into a cohesive whole was one concern in the mid-1990s. Another task to be addressed was the relationship with Boeing. Rockwell Collins equipment had important standard positions on several Boeing aircraft before being displaced by Honeywell on the 777. In some cases, however, those earlier aircraft programs had experienced cost overruns and delays that led the customer to question the Collins Air Transport Division's performance.

In some views, those problems also kept the division from investing enough in developing the capability to provide new-technology LCDs for cockpit instrumentation and furthering its efforts in integrated avionics. Missing out on the 777 programs — which Honeywell won with a proposal that included LCDs — also led to losing out on the lucrative 737 Next Generation. It was especially painful because Rockwell Collins equipment had virtually all of the standard avionics and displays positions on the earlier Boeing 737. Collins Commercial Avionics had bet that Boeing would pass on the then-risky new display technology in both the 777 and 737NG programs. They offered what they were prepared to deliver: reliable but older and heavier cathode ray tube (CRT) displays.

Boeing, however, was committed to LCDs on the 777. When the division later proposed using CRTs on the 737NG and upgrading in the future, a top Boeing executive later explained: "We could ill afford to orphan the queen of the fleet by taking a step back in technology in the 737."

The company was on the outside looking in with its biggest commercial customer. The organization would spend much of the next decade rebuilding that relationship. As Jones noted, memories in the commercial aircraft industry are long, and winning back Boeing's trust would not be easy: "Once you get out of bed with them, it takes a long time to get back in. They don't just turn around and say, OK, you're a good guy, here's all my business."

› Liquid crystal displays (LCDs) used in this Pro Line 21 system and others showed complex information graphically and became a standard for new aircraft.

"My charter when I came here was to make this a more customer-focused, responsive organization," Jones said later. Before, it had been perceived as — and, in fact, was — an engineering-centric organization, one that was built around engineering excellence and innovation. That reputation and position had served the Collins businesses well for the most part, Jones noted.

A first step in becoming more customer-focused was changing customers' impressions of the company. For example, the flagship Building 124 on the west side of C Avenue traditionally had housed engineering, which often meant that meetings with visiting customers were held elsewhere on the "campus." That was changed.

To be seen as a world-class company, Jones said, "You have to look like a world-class company. We may not have always thought like a world-class company in terms of how we positioned the company and what we did in terms of growing the company."

As right-hand man to Jack Cosgrove when the companies were merged, Jones set about raising expectations for what the company could achieve. That applied to employee expectations as well as those of customers. He addressed the Boeing issue head-on and early, letting the company know that Rockwell Collins was eager to demonstrate that it could, and would, deliver on existing programs and "reaffirm their trust."

> Rockwell Collins' commitment to LCD development was followed by a string of successes with Boeing and other aircraft manufacturers.

"For a lot of reasons, we had to raise our credibility," Jones recalled. "The strategy was to go there and get a clean slate."

Before long, A&C took an important step back into Boeing's good graces by making a commitment to being a second displays provider behind Honeywell, which wasn't performing to Boeing's satisfaction on the 777 and 737NG. While Honeywell effectively blocked the Collins participation at that point, the commitment ultimately paid off. In addition to improving performance, it took a bit of brinksmanship with Boeing to get the relationship reestablished.

Boeing desperately needed more than one displays provider to ensure that it could deliver on new aircraft orders and upgrades. A&C had rejuvenated its investment in LCD technology but needed to know if Boeing would commit to being a customer.

Jones told Boeing development executives, "We have this display, and we're either going to bring it forth to allow you to select it, or we're shutting it down. If you're telling me we're not in this business, there's no reason for us to continue to invest in it. It's a waste of our time."

When displays were upgraded in Boeing's 767-400, they were provided by Rockwell Collins Avionics & Communications. In mid-1998, Boeing signed off on the systems/hardware critical design review for the Large Format Display System (LFDS), clearing the way for Rockwell Collins to proceed.

"We've never lost a Boeing display contract since," Jones said in 2009. Significantly, displays technology was seen as so important to the company's future that the first of the Rockwell Collins "Centers of Excellence" was formed in 1997 to address it. The centers focus on refining key technologies for use across the businesses, from large air transports to business and regional jets to military aircraft and vehicles.

Advanced, interactive displays

Understanding human factors and how humans could most effectively interact with computers was becoming an increasingly important issue in the mid-1990s. A team led by the Collins Avionics and Communications Division helped put Rockwell Collins in the midst of those developments when it won a five-year, $25 million research project from the Army Research Laboratory (ARL) to study human-computer interactions. The ARL project covered issues such as computer vision, image and video compression, eye movement and human perception, visualization, design and tactile communications.

That program and later work in the Displays Center of Excellence led to patented developments such as anti-reflective and anti-glare display glass as well as touchscreen capabilities for avionics. Advanced lamination processes also helped sharpen the visual displays, which were optimized for clear viewing and readability over a much larger viewing angle than was possible with early LCD technology.

The Displays Center of Excellence was established in November 1997 and brought together some 100 technical experts and business leaders from the Advanced Technology Center and the Rockwell Collins divisions. Jerry Gaspar, who was the center's first vice president, explained the rationale for the initiative: "Historically, the separation of the divisions has caused the company to produce many very similar displays with very different architectures and components. The Center will consolidate the designs in the displays area so we avoid needless duplication. That way, we'll get to market faster while lowering our development and manufacturing costs."

Reuse of those designs was expected to simultaneously benefit all three key Rockwell Collins markets: business and regional aviation, air transport, and military, particularly with military buyers pushing to find more and more off-the-shelf solutions. ▪

› The Displays Center of Excellence was responsible for patented developments in anti-reflective and anti-glare glass, touchscreen capabilities and an advanced lamination process that helped sharpen visual displays.

The "Lean" journey

Rockwell Collins got more than business from Boeing. After Clay Jones and Herm Reininga participated in an all-day supplier conference on the concept, the company also emulated the aircraft giant's adoption of Lean manufacturing. They took the Lean concepts of continuous improvement, eliminating waste and striving for perfection and applied those concepts not only to manufacturing, but to the entire enterprise.

The timing of Jones' exposure to Boeing's Lean manufacturing initiative was an example of the "serendipity" that he would later say is characteristic of his career.

"We had a sort of quality issue that we couldn't make go away," he explained. "Products were improving, but not fast enough. Development programs were improving, but they weren't improving fast enough. I thought we needed a bellwether program, a program we could stay with that had a half-life greater than a year. I didn't know what it was."

› Lean Electronics became a tool for changing the Rockwell Collins culture.

It turned out to be a customized version of Lean manufacturing, a first in the industry, which was named and even registered as a service mark: Lean Electronics℠. It was to become a tool for changing the culture of the Rockwell Collins business as well as improving processes and quality — a means to achieving the goal of operational excellence. As with many previous improvement programs, "Lean" was met with an undercurrent of cynicism by some in the organization, Jones said. There was a concern that it was just another "initiative du jour" that would last a year or so only to be replaced by another.

"It took a long time for people to realize it isn't going to go away," Jones recalled. "We had some people, as we always do, who are early adapters and rushed to the concept. We had resisters. We had resisters, frankly, at the senior level."

A new Lean Electronics organization was established to integrate efforts focused on building the core business, expanding into new products and reaching new customers, and achieving operational excellence.

> Lean Electronics initiatives touched virtually every aspect of doing business at Rockwell Collins and influenced suppliers and customers.

> Employees participated in RPI workshops to identify and eliminate inefficiencies and waste.

Over the next decade, several different initiatives took place under the "Lean" umbrella. They would include Project Fusion, a Core Process Optimization initiative, and establishment of various tools such as Radical Process Improvement (RPI) workshops. Numerous RPI workshops were organized around the concept that process improvement should be radical and quick — 80 percent complete by the end of each workshop. Teams were encouraged to take risks as they worked to strip away waste and create streamlined, agile, value-added processes. Hundreds of RPIs were held in the first few years of Lean as virtually every process was closely examined and updated.

The advent and success of Lean Electronics inspired a similar effort at the corporate level, where an initiative known as Rockwell Lean Enterprise was established with radical goals that are characteristic of Lean thinking: 50 percent reductions in cycle time and inventory, 25 percent reduction in floor space and 30 percent reduction in costs and defects over four to six years.

Sharing of the Lean philosophy was also extended outside of Rockwell International and Rockwell Collins — to suppliers and customers as well. Lean Electronics workshops were held to help parties on both ends of the spectrum reduce their waste and improve performance.

Partnering, divesting, acquiring

The same year that brought the Collins businesses back together, 1996, Rockwell Collins and Marconi Electronic Systems — a longtime arch-rival and occasional business partner — formed a limited liability company called Data Link Solutions (DLS). The Collins businesses and Marconi, which later merged with British Aerospace to form BAE Systems, were longtime competitors for the government's JTIDS and MIDS programs and others.

DLS formalized a relationship that Kelly Ortberg would later say was "far and away the most successful joint venture" for the Collins business. After initially participating as a subcontractor in JTIDS development because of its expertise and reputation as "a great radio company," he said, Rockwell Collins became "a total network communication provider."

Also in 1996, a team that included CACD and Honeywell, another arch-rival, as a companion subcontractor was recognized with two prestigious McDonnell Douglas Corporation awards for its role as a supplier to the McDonnell Douglas Aerospace (MDA) division on the Joint Direct Attack Munitions (JDAM) program. In April 1996, CACD received the MDA Model of Excellence award, then went on to qualify for and win the corporation's Spirit of Excellence Award.

The JDAM program was one of several Pentagon acquisition-reform pilot programs. The JDAM guidance system, for which CACD provided GPS technology, was devised to improve the accuracy of so-called "dumb" bombs. The success of the program was attributed in part to the teaming relationship established under MDA's Integrated Product Team (IPT) framework. It required extensive cooperation between normally competing companies.

When the MDA recognition awards were announced, Bob Chiusano said, "You know the IPT is truly working when you look in the room and watch people from several corporations working together, and the boundaries between the companies are lifted, and it's impossible to tell who works for who, because they're all focused on winning."

Al Caslavka, then the JDAM program manager for CACD, noted, "CACD went from not being a preferred supplier [to MDA] to attaining a silver level of preferred supplier certification within 12 months."

› The DLS joint venture brought together traditional competitors to grow in the network communications market.

› Left: An F-15 Eagle drops MK-84 Joint Direct Attack Munitions over the Navy's China Lake test range near Edwards Air Force Base in California.

U.S. Air Force Photo

› JDAM guidance systems made "dumb" bombs more accurate by using GPS technology.

Coincidentally, 1996 also was the year that the major Rockwell Collins businesses established their official presence on the Information Highway. Rockwell International had begun using the Internet for marketing purposes in 1994 and was urging its various businesses to follow suit. Despite their growing influence in avionics network communication, however, Collins Commercial Avionics and CACD did not venture beyond using an internal Bulletin Board System to engage on the World Wide Web until February (CCA) and March (CACD) of 1996.

It had become clear that not being on the Internet could give potential customers the unfortunate impression that the high-technology company wasn't using technology effectively.

The nature of the avionics business and a sensitivity to protecting intellectual property meant the Rockwell Collins units had to take special care when deciding what information to make available. Not only would customers have access to information on the Internet, but so would competitors and foreign powers. Planners had to find a balance between giving information to customers and giving information away to competitors. Early on, service bulletins were made available to customers, but protected by passwords. Product specifications were published but soon removed from the online parts catalog.

The Rockwell Collins businesses had operated across much of the globe for decades, and had engineered sophisticated systems that allowed AWACS aircraft, fighter planes and ground installations to share data across high-speed networks. Now, they were sharing some of their own information with the world in the ultimate network.

Much as Rockwell International had sold or spun off some of its businesses, the new Rockwell Collins set about divesting or shutting down business efforts that were considered "non-core." The Railroad Electronics division was sold in May 1998 to Westinghouse Air Brake Company. Forays into GPS and other electronics for agriculture and local mass transit systems were also ended. A new strategic focus on expanding the core avionics business was taking hold.

> Acquisitions of Hughes-Avicom International and Sony Trans Com (below) firmly established Rockwell Collins as a provider of in-flight entertainment systems.

The previous year, Rockwell Collins had entered the in-flight entertainment (IFE) business by acquiring Hughes-Avicom International, a subsidiary of Hughes Electronics Corp. The additional purchase of Sony Trans Com in 2000 further expanded the company's IFE capabilities under the name Rockwell Collins Passenger Systems. The company also acquired a Cedar Rapids-based avionics parts supplier, Intertrade, Ltd., in 1999.

To further establish its role in providing head-up display technology, Rockwell Collins in 1999 acquired Kaiser's share of Flight Dynamics. The company was the only supplier of head-up displays in the regional aircraft market and a leader in the commercial air transport, business jet and government air transport markets. The Flight Dynamics Head-Up Guidance System® (HGS) displayed critical flight

› Kent Statler served in several top management roles after joining the company in 1987.

› A maintenance technician works on avionics gear accessible in the "hell hole" of a Boeing 757. As airlines sought ways to reduce in-house costs, long-term maintenance agreements became common.

Photo by Russell Munson

Getting paid for success

While Rockwell Collins was establishing or solidifying core areas of expertise in advanced, integrated avionics and displays, its commercial division was reexamining its maintenance relationships with its end-user customers, the airlines. In the midst of a 1990s recession in the air transport industry, the operators were looking for ways to reduce their own, in-house maintenance and engineering costs, which were estimated to account for as much as 25 percent of their direct operating costs. Collins Commercial Avionics commissioned a study by Arthur Little and Company to survey industry attitudes about outsourcing of certain maintenance functions. It found there was strong interest in contracting out the responsibility for maintaining avionics systems and equipment.

The survey results led to the industry's first-ever "per-flight-hour" avionics maintenance program, a sort of partnership between an airline and the avionics provider. Kent Statler, who was named executive vice president of Rockwell Collins Services in 2005, said the company essentially takes over the management of maintenance and related logistics. While previously the customer maintained an inventory of spare parts, Rockwell Collins Services ensures that parts are available when and where they're needed to limit expensive ground delays for the airlines. Instead of paying steep repair costs and having aircraft out of service, the airlines began to pay a contracted rate for every hour the planes are able to fly. American Airlines and TWA were among the first to participate in the program by signing long-term contracts.

Statler said the innovative agreements replaced "getting paid for failure" when equipment needed expensive repairs with "getting paid for success" by keeping aircraft in the air and making money for the customer. ▪

› Rockwell Collins' acquisition of K Systems helped solidify the company's position as a provider of head-up displays, including the Flight Dynamics Head-Up Guidance System.

Photo by Russell Munson.

information superimposed on a clear transparent glass screen in the pilot's forward field of view to aid in making more precise, safer landings.

In 2000, Rockwell Collins acquired Kaiser's parent company — K Systems — strengthening its display capabilities for military applications and filling a void in the Rockwell Collins product portfolio.

Pro Line 21 spurs business aircraft successes

While CCA had not been far enough along in LCD development to snare important Boeing contracts in the early 1990s, its General Aviation Division grabbed a lot of attention for employing that very technology with the introduction of the Pro Line 21 avionics suite in May of 1995. Positioned as "21st Century" avionics, it was received enthusiastically by aircraft manufacturers and their customers, as well as by industry media. *Business & Commercial Aviation* magazine proclaimed Pro Line 21 as "the most powerful avionics product line in business aircraft history."

With the integration of a new Collins flight management system, it also led to tremendous growth in market penetration for the division, which in the mid-1990s became known as Business and Regional Systems. Steve Belland, Pro Line 21 program manager who later became Rockwell Collins vice president for acquisitions, said it ultimately turned Cessna into a major customer. From purchasing roughly $2 million annually in Collins radios, and most other avionics from Honeywell, Cessna became a nearly exclusive Rockwell Collins avionics customer with buys worth tens of millions of dollars annually.

"Once we really started focusing on human factors and bringing in the flight management system, we started winning a lot of positions," Belland recalled. "When we came in with Pro Line 21, it absolutely shocked everybody."

The system included large, easily readable LCD displays, the Collins AVSAT™ satellite-based communication and navigation system, and top-quality GPS receivers,

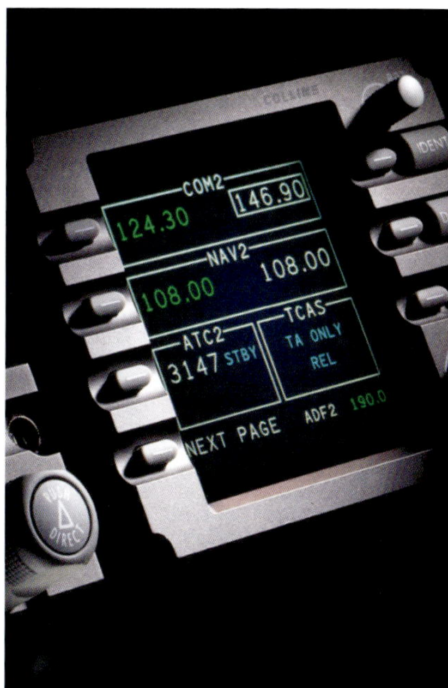

› Left: The complete Pro Line 21 avionics suite featured a head-up guidance system, five 8-by-10-inch LCDs, two flight management system control display units and autopilot and communications controls.

› Above: The RTU-4200 digital radio tuning unit.

along with other "human factors" improvements worked out with extensive customer input. It was designed to readily adapt for use in a variety of cockpits and could be configured with two, three, four or five adaptive displays — adaptive because of their context-sensitive formats, which allowed pilots to see the most appropriate information during each phase of flight. Altitude, airspeed, attitude, engine data, navigation maps, TCAS and radar information could be positioned on a single display. The new suite also offered the capability of using miniature digital cameras mounted to allow the pilots to check landing gear or ice build-up on wing surfaces.

Not only did Pro Line 21 include new electronics, it was developed in just a year rather than the previously typical three years. Compressing that time frame hinged on involving end users early in the process, and letting them "fly" with a simulation system to test out new designs and changes before software code was written. A concept in 1995, the Pro Line 21 system underwent its first actual flight test in August 1996 aboard the company's Sabreliner test plane.

By 1999, the division had captured new integrated avionics programs for both the Cessna Citationjet I and Citationjet II programs and the Bombardier Continental mid-size jet. Rockwell Collins was well-positioned to grow in both the business jet market and the market for regional jets, for which it also developed and integrated a Pro Line 21 flight management system. Belland noted that the company also benefited from increasing segmentation in business jets, as growing numbers of wealthy individuals joined corporate buyers to drive market growth.

KC-135, GATM program growth continues

As the commercial side made inroads in the business and regional airline markets, the company's government business continued to grow, driven partly by a timely weather radar improvement. In 1996, Rockwell Collins delivered the first KC-135 tanker aircraft upgraded in the Pacer CRAG program, which had been awarded in October of the previous year. Those and later KC-135 aircraft were among the first to use the Collins

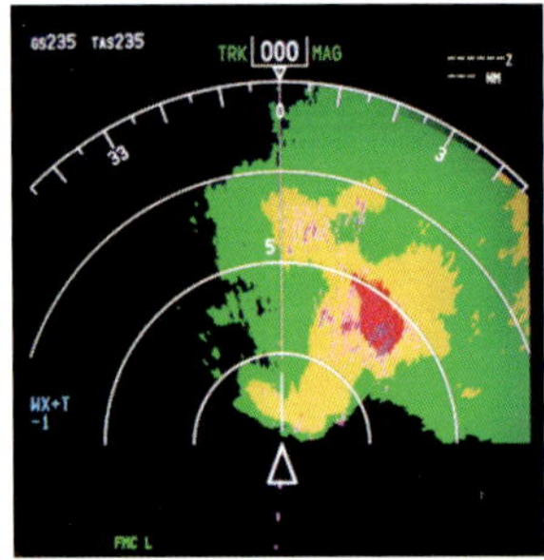

› The WXR-700 weather radar system activated automatically during takeoff and landing to give pilots critical advance notice of wind shear conditions.

Forward-Looking Windshear Radar system. (It was installed first on a Boeing 737 owned by the Orlando Magic NBA team.)

The upgrade to the WXR-700 air transport radar was developed to give pilots advance warning of wind shear conditions, which can subject aircraft to sudden, violent and potentially deadly downdrafts. Previous wind shear detection devices notified pilots only after they were already engulfed in wind shear and provided navigational advice for escaping.

NASA studies had shown that even a 10-second warning could be enough to allow a pilot to avoid the wind shear or better position the aircraft to pull through it. The new Rockwell Collins radar gave pilots a minimum of 10 seconds' warning and as much as 60 seconds. It was designed to activate automatically during takeoff and landing, scanning the area ahead of the plane for microburst wind forces. Certification requirements were particularly challenging. The Air Transport Systems Radar Team used exhausting simulations of a variety of wind shear events and had to prove the system could perform in actual wind shear conditions.

After having limited success placing the WXR-700 with military customers, the addition of forward-looking wind shear capability was seen as a major breakthrough that helped place the Collins weather radar in the KC-135 fleet. That program also included the addition of a "skin paint" function that displays the location of other aircraft on an LCD screen to help tanker pilots rendezvous for refueling or flying in formation with others.

In a subsequent program, the company won a $600 million contract to upgrade hundreds of KC-135s to meet new Global Air Traffic Management (GATM) requirements for military aircraft to operate in civil airspace. The earlier successful KC-135 program had helped demonstrate that Rockwell Collins could deliver. Also contributing to that follow-on program win was the development, beginning in 1997, of a next-generation open systems architecture (OSA) that was designed to serve the air transport, general aviation and military markets.

› The KC-135 flight deck was updated with Rockwell Collins avionics systems to meet GATM requirements.

The OSA effort was begun when the company recognized that using a common approach would lower development costs and get new products to market faster. The OSA concept's initial impact was seen on fixed-wing aircraft. It was later expanded and applied to numerous platforms and programs, including rotary-wing aircraft, ground systems and other aircraft.

Selected as integrator of the KC-135's flight deck systems in October 1999, Rockwell Collins upgraded the aircraft to meet new Communications, Navigation, Surveillance/Global Air Traffic Management (CNS/GATM) requirements to enable them to operate in both commercial and military airspace in a developing new "free flight" environment.

As with the earlier Pacer CRAG program, the KC-135 GATM upgrade made use of some of the latest commercial, off-the-shelf (COTS) solutions for communication and navigation equipment. Major components included the RT-1341(V) 8 High Frequency (HF) radio, GNLU-955M Multi-Mode Receiver (MMR), VDL-2000 VHF data link capability, SAT-2000 SATCOM radio, CMU-900 Communications Management Unit and the IPC-7000 Integrated Processing Center. (The MMR landing system was already in use in civilian aircraft around the world. News of its selection by both Airbus and Boeing had been announced at the 1996 Farnborough Air Show near London.)

› A pair of Royal Air Force GR4 Tornado fighter aircraft approach a U.S. Air Force KC-135R Stratotanker to refuel.

› The IPC-7000 (above) is a modular processing and data networking system that can host multiple mission, flight management or display management processing functions while providing extensive functional growth for additional applications.

A program win in 1998 with FAA-certified COTS "free flight" products was considered one of the first big payoffs from the consolidation of the Rockwell Collins commercial and government businesses. Boeing had sought a commercial supplier with an understanding of military needs to outfit its C-17 airlifter. Air Transport and Government Systems personnel worked together to win a $12.4 million award, putting the team in a position to capture significant additional military aircraft programs — including the KC-135, KC-10 and C-130 — and to become the major player in GATM avionics for the military.

Rockwell Collins won another GATM position in 1998 with a $12 million contract from the Naval Air Systems Command to supply its military ARC-210 hardware for the C-5 Galaxy cargo aircraft. It suffered a setback early in 1999, however, when the Air Force chose a higher-priced but more integrated, commercial-based GATM system for the C-5. Executives recognized the loss as an important wake-up call, which resulted in still closer teamwork between Government Systems, Air Transport Systems, and Business and Regional Systems, and ultimately led to the successful KC-135 program.

For the C-17, Rockwell Collins was selected to provide its SAT-2000 system for airborne satellite communications, along with the CMU-900, a communication management unit data router that transmits and receives messages through VHF, SATCOM and HF data link channels. In a cooperative arrangement, Government Systems acquired the systems from Air Transport to provide them to Boeing. Both sides of the business had invested in the development costs.

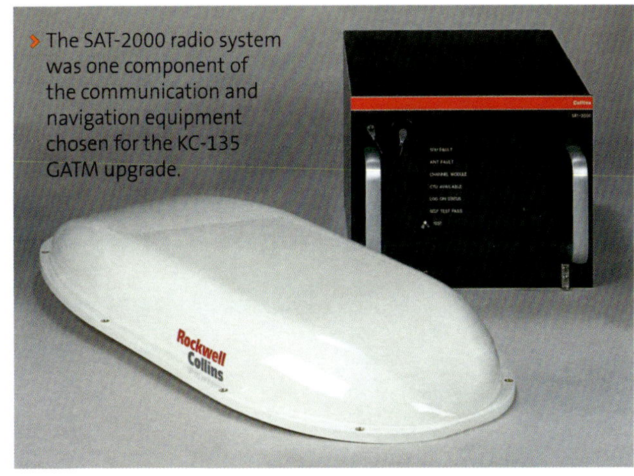

› The SAT-2000 radio system was one component of the communication and navigation equipment chosen for the KC-135 GATM upgrade.

FORCE

‹ The Air Force's C-17 carries troops and equipment and uses Rockwell Collins satellite communications and data links systems.

By accepting a COTS offer without traditional military requirements, Boeing reduced its costs for the Collins systems by an estimated 25 to 50 percent. The more streamlined acquisition process was likened to going to a computer store and selecting a computer "off the shelf" that would meet the buyer's needs. Savings to taxpayers were expected to be substantial as the new way of doing business became more common.

› A&C partnered with Chinese firms to market GPS receivers and vehicle tracking systems.

International successes

Within a few years of Jim Churchill's retirement, more of the international growth he had hoped to see was being realized, notably in such previously difficult markets as Russia and China.

The Collins Traffic Collision and Avoidance System (TCAS) became the first TCAS certified in Russia in 1996. It was in use by more than 110 airlines worldwide at that point. The following year, its operating range would be expanded to more than 100 nautical miles – in excess of twice its previous range.

Rockwell Collins also successfully demonstrated satellite-based air traffic management (ATM) systems in both China and Russia in 1996, and won another contract to conduct ATM trials in Saudi Arabia. The company also established a new venture in China to market commercial GPS products as Shanghai Rockwell Collins Navigation and Communications Equipment Co., Ltd.

A&C's partnership with two firms owned by the Chinese government — the Shanghai Avionics Corporation and the Shanghai Broadcast Equipment factory — was created to build, sell and service maritime GPS receivers and vehicle tracking systems in the rapidly expanding Chinese market. It prompted a visit to Cedar Rapids in November 1996 by Zhou Wenzhong, minister of the Embassy of the People's Republic of China. By then, Rockwell Collins was a major supplier to many of China's regional, domestic and international airlines.

Workers in the Chinese facilities had actually begun to manufacture Collins flight instrument and weather radar systems under licensing agreements in 1989. To continue to grow in China, establishing a local presence that would produce local content was seen as a key. The Collins Communication Systems Division was working closely with the Chinese Research Institute of Navigation Technology to develop and field-test a new satellite-based air traffic management system. CACD was also pursuing business in China, introducing new HF voice and data communications systems.

Other commercial business efforts resulted in additional penetration in overseas markets in 1997. Air China, that country's largest international carrier, selected the Collins MMR and Series 900 communication and navigation sensors for its new Boeing 777 aircraft. Aeroflot selected advanced Collins avionics for 10 new Boeing 737-400 aircraft. At the Paris Air Show, Air France announced Rockwell Collins' selection as its preferred avionics supplier for its current fleet and new aircraft. A five-year agreement called for communications, navigation, collision avoidance and surveillance avionics and other Rockwell Collins avionics to be installed on the airline's new Airbus A-320 and Boeing 777 aircraft.

› Air France was one of hundreds of airlines worldwide using Rockwell Collins avionics.

> The 100,000th PLGR unit was delivered to the military in 1997.

Years of recognition, milestones

The late 1990s were particularly eventful for Rockwell Collins, which was recognized as one of the nation's leading manufacturers. Following that honor were a number of successful new product launches and the achievement of several notable technological and production milestones.

In October 1996, *Industry Week* listed the Cedar Rapids and Coralville plants as among the top 10 in the United States. In an article about the selection, the publication cited a 71 percent increase in productivity over the previous five years, a 21 percent reduction in per-unit manufacturing costs and a 55 percent reduction in the product development cycle, among many other factors. It attributed much of the improvement to the creation of self-directed work teams that resulted from the unique cooperation between management and International Brotherhood of Electrical Workers (IBEW) Locals 1634 (Coralville) and 1362 (Cedar Rapids).

Moe Kassem, then IBEW Local 1634's business manager/president, was quoted as saying: "In this partnership, we have to respect the different areas of responsibility and realize that the competition is with the outside.... I want Rockwell Collins people to be the best, and I believe that the union cannot be the best without management excelling in organizational leadership."

Plant manager Kent Statler explained in the article what made the partnership work: "Joint partnerships call for mutual respect and mutual understanding. On either side, be prepared to appreciate that it is impossible to please all the people all of the time. Above all, I can't overemphasize the importance of communicating."

> The small, lightweight AHS-3000 attitude/heading reference system (left) used digital quartz MEMS sensors to give pilots essential heading and attitude information. It replaced much larger, mechanical gyroscopes (right).

The Coralville plant passed another significant milestone in GPS development when it delivered the 100,000th PLGR unit to the military in 1997.

July 1997 also brought the announcement that Don Beall — who had served as president of Collins Radio Group in the 1970s — would be stepping down as Rockwell's chairman and CEO. He had held those posts since 1988 and was succeeded by Don Davis. Beall later served as the first chairman of the Rockwell Collins board of directors and chairman of the board's executive committee.

Continued growth in sales illustrated Rockwell Collins' increasing impact on the overall Rockwell International picture as well as the corporation's changing strategic focus. Rockwell Collins A&C sales grew 15 percent in 1997 to $1.7 billion, when the corporation's overall sales totaled $7.8 billion. After selling its Aerospace and Defense businesses to Boeing, only 7 percent of Rockwell International's sales were to the U.S. government in 1997.

The Collins AHS-3000 was one of four key new Rockwell Collins products introduced that year. A "next generation approach" to attitude/heading reference systems, it used digital quartz micro electro-mechanical systems (MEMS) sensors to provide critical heading and attitude information to aircraft flight decks and other on-board systems.

The new SATCOM-5000 and 6000 satellite communications systems extended global voice, data and fax capabilities to business aircraft flight decks and cabins. The Collins APR-4000 GPS approach sensor was acclaimed for offering precision approach capability. The TCAS-4000 was notable for doubling the surveillance range of existing systems to more than 100 nautical miles, which would be critical for supporting free flight operations. That global initiative to liberate airlines from following mandated, often indirect and inefficient routes at specified altitudes was seen as a potentially huge market for GPS technology that would save millions of dollars in fuel costs.

In 1998, Rockwell Collins captured a record $1 billion in orders for government systems and ended the year with a record backlog. In September of that year, Boeing and Rockwell Collins celebrated the successful first flight of Boeing's new business jet, which carried an entire suite of Rockwell Collins communication, navigation and surveillance equipment.

While relations between labor and management generally had been good throughout the history of the Collins businesses, members of the Coralville local went on strike on May 28, 1998. Bargaining unit employees in Cedar Rapids and Dallas had ratified new five-year contracts. Coralville employees returned to work 16 days later when a new agreement was reached.

Other notable developments came in various segments of the military and commercial markets: The company was selected to provide data link upgrades for 52 Airbus aircraft on order for United Airlines; the ARC-210 Advanced Communication System was expanded for use across all U.S. military and Allied Forces airborne platforms; and the government ordered Rockwell Collins avionics upgrades for installation of a glass cockpit on the Sikorsky Blackhawk helicopter.

Rockwell Collins also was awarded a standard position for autopilot flight controls on the Boeing 737NG line, and the flight deck large-format displays for the 767-400, which was the first aircraft in the industry to use Rockwell Collins applications of Ethernet technology.

› The TCAS-4000 doubled the range of earlier traffic collision and avoidance systems.

› Rockwell Collins provided avionics upgrades for Blackhawk helicopter glass cockpits.

The Lean Electronics "journey" was well under way in 1998, when some 35 RPI workshops were completed, resulting in major reductions in inventory levels, a 22 percent reduction in floor space requirements and a 30 percent improvement in productivity. By mid-1999, the total had grown to nearly 150. Hundreds more were scheduled through 2000.

One goal had been to cut costs through Lean Electronics by 35 percent, Jack Cosgrove recalled. Between 1996 and 1999, when he retired, the company had achieved that goal "and then some," he said.

› Clay Jones succeeded Jack Cosgrove in 1999.

A part of "something special"

When he was named Avionics & Communications president in 1996, Cosgrove became only the third person to preside over a unified Collins company. He followed Arthur Collins and Robert Wilson, who was president of Collins Radio Company from 1971 to 1973. When Cosgrove retired in 1999, he was succeeded by Clay Jones.

In an interview published in Rockwell Collins' *Millennium* Magazine soon after he became president, Jones was asked if there were similarities between running a $2 billion business and piloting an F-15, which he had done in the Air Force. He dismissed the notion, but said being in the military had helped prepare him in some ways.

"Obviously when you're flying at 600 to 700 miles an hour, you have to make quick decisions. I also think it breeds self-confidence and teamwork. And finally, I think being a fighter pilot teaches you to be, if you're not already, intensely competitive. I love to win, and hate to lose — and I think I bring that same competitive spirit to this job as well."

He also said Cosgrove had taught him "the benefit and the art of patience, and the value of having the patience to listen to other points of view before you make a decision."

Ten years later, Jones said his relative newness in the Rockwell Collins organization was an advantage, considering the task at hand: his charter to turn the focus onto the customer.

"All of my experiences at Rockwell International had been focused on the customer," he said of his previous 15 years in marketing and business development. He was used to thinking like a customer and talking to customers on a level that would help him understand their needs and how the reorganized company could meet them.

"I wasn't very long in the tooth in this business," he said. "If there was anything that was probably working in my favor, it was the fact that I had very little corporate memory, so I didn't know how we'd always done it."

In January 1999, shortly after he became president, Jones had a pivotal meeting with his senior leadership council. It was the second year of "Lean" and some were still not on board. He made what he later called an "impassioned" speech to spur on the reexamination and refining of the "core" processes he was convinced would fuel the company's future.

He challenged those present to be part of "something special" — to be more concerned about making the company successful than about losing their jobs.

"Everybody came around," he said later. "That cemented the consolidation." ▪

Late in 1999, the company established its Decorah production facility as a "model" Lean Electronics facility after it achieved significant improvements in productivity and quality. It was used as an example to lead Lean transformations throughout the Rockwell Collins manufacturing facilities.

The decade, and the century, ended with still more growth for Rockwell Collins despite a sizeable dip in the production of new air transport aircraft. Rockwell Collins' sales grew by 5 percent in 2000, a year also marked by a number of new Air Transport Systems product introductions, including the first iteration of its MultiScan™ weather radar.

The company's continued success was driven in part by overseas in-flight entertainment sales as well as ongoing success with Pro Line products for business and regional aircraft, a new emphasis on service, and software for wireless military communications.

A revolution in weather radar technology

The introduction of the MultiScan Weather Radar System in 2000 for air transport customers represented a whole new way of thinking about weather radar. What had been a time- and attention-consuming task that pilots mastered only with extensive training and years of experience became a more reliable, simpler and nearly hands-free matter of routine.

The MultiScan system automatically scanned ahead of the aircraft to detect weather out to more than 300 nautical miles. It combined the radar returns through advanced digital processing and analysis algorithms to display not just precipitation rates but actual weather threats so pilots could take action to avoid them.

Over the next several years, Rockwell Collins engineers continued to refine the system, incorporating input from international airlines and even undertaking an historic, worldwide research and development effort. After the product's initial launch on a Boeing 747, principal systems engineers Dan Woodell and Roy Robertson learned that the radar's performance could vary greatly in certain geographical areas. Research by Dr. Edward Zipser at the University of Utah showed that dramatic differences in regional weather patterns would require multiple weather detection solutions.

In 2007, with a multimillion-dollar funding commitment in hand, the MultiScan team leased a Boeing Business Jet and began a series of more than 30 test flights to analyze weather phenomena in the Asia-Pacific region, North and South America, Africa, the Caribbean and the North and South Atlantic. The resulting MultiScan Hazard Detection System helped pilots around the globe make better decisions and improved flight safety, comfort and efficiency.

MultiScan weather radar systems were recognized with *Aviation Week's* Technology Breakthrough Award (2004), the Airline Avionics Institute's Volare Award (2005), and *Flight International's* award in 2007 for best product in the Propulsion, Systems and Avionics category.

The MultiScan system proved so successful that it helped sell other Rockwell Collins systems. "The vast majority of the times when we compete in this arena, we end up winning contracts with this radar," Woodell said. The system was certified in virtually all Boeing and Airbus platforms in production and also was adapted for the business and regional aircraft markets. ▪

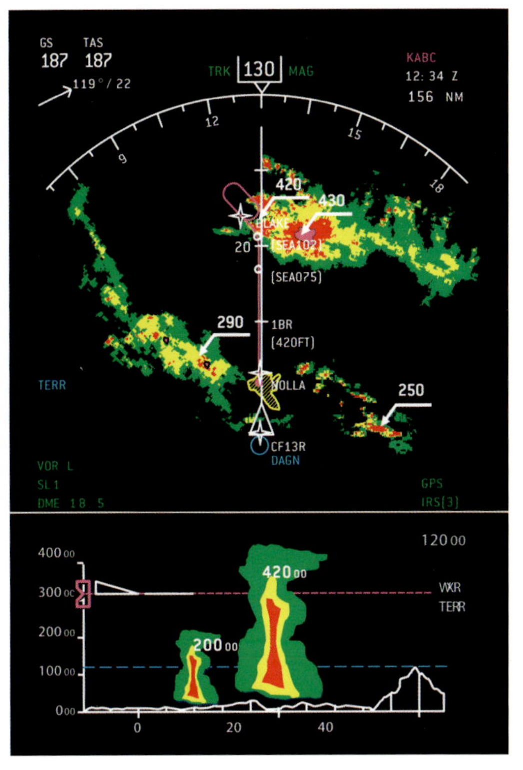

› A Rockwell Collins MultiScan weather radar display.

› The Total Entertainment System (TES) brought video programming to airline passengers in their seats.

Qantas Airways, for example, selected Passenger Systems' Total Entertainment System (TES) for its fleet of international long-haul Boeing 747-400s and Scandinavian Airline Systems selected TES for its Airbus 330 and A340 aircraft. Passenger Systems also received significant follow-on orders from American Airlines, Air France, British Airways and Japan Airlines.

Business and Regional Systems ended 2000 with record sales, which included Bombardier Aerospace's selection of Collins Pro Line 4 avionics for its new Canadair Regional Jet Series 900. It also began production deliveries of Pro Line 21 on Cessna's Citation CJ1 and received Pro Line 21 certification for the Citation CJ2.

Collins Aviation Services, a relatively new business focus, won strategic service business with Express Airlines One as well as with the U.S. Navy for servicing ARC-210 communication systems. It also developed an industry-leading service network and

AUDIO JACK
A/T DISC
A/T DISC

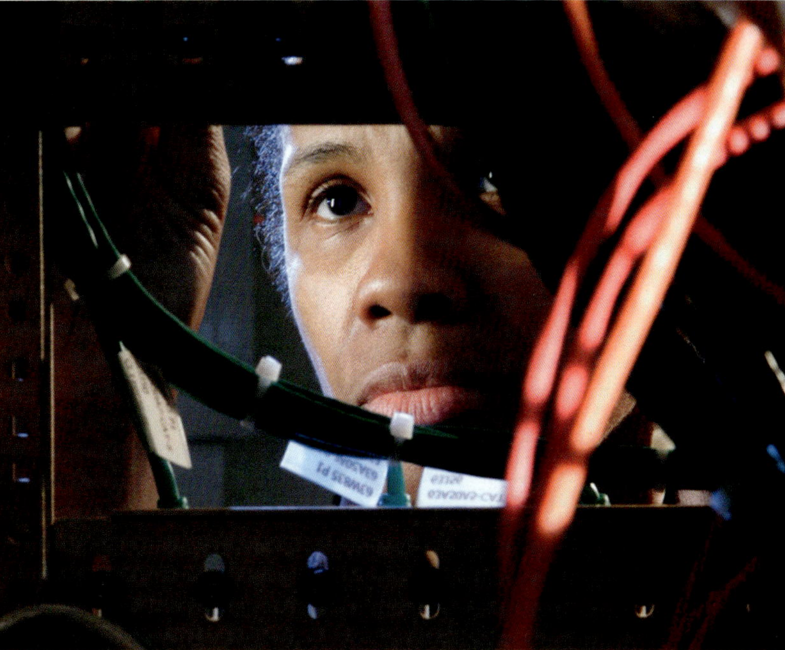

announced new alliances with BFGoodrich Aerospace, FLS Aerospace and Hamilton Sundstrand for aircraft component service.

Rockwell Collins Government Systems introduced an advanced HF data communications tool to provide wireless messaging for the military. The HF Messenger™ software was developed for ground, tactical airborne and maritime users and selected in 2000 by the U.S. Army Communications-Electronics Command and by the U.S. Navy for its Battle Force E-Mail System.

Year 2000 also marked the start of a new Rockwell Collins "eBusiness" organization to coordinate existing information technology and Enterprise Resource Planning efforts and to create two new entities: eServices to integrate services electronically with customers and business partners, and Lean Services to provide strategic alignment and continue to drive operational efficiency.

> More women and minorities joined the Rockwell Collins workforce as the company expanded internationally and culturally.

Embracing diversity

Along with the rest of society, the engineers, executives and factory workers of Rockwell Collins had experienced seemingly one clash of cultures after another every few years.

In the early 1980s came software engineers, who were quite different from those who had designed and built Collins radios. Business development executives learned that they couldn't show up in business suits and easily sell GPS systems to railroaders and agriculture-industry buyers who wore seed caps. Some learned to operate in Europe, Asia and other parts of the world where business practices and the speed of decision-making could be a far cry from what they were in North America. Rockwell Collins had become far more than a U.S. company, with hundreds of employees in Mexico, France, Great Britain, Russia, China and Australia.

Until the early 1990s, the engineers recognized for accomplishments in the pages of the *Rockwell News* were almost exclusively male. The few women pictured were either factory workers or in clerical positions, being recognized for achieving certification in secretarial skills.

By the mid-1990s, however, there was a growing recognition in the Rockwell Collins world that diversity was going to play an ever-larger role in business, as well as in other parts of people's lives. More women were arriving with engineering degrees, and the workforce was becoming more international.

Don Beall had addressed how diversity related to the corporate vision as early as 1992: "The challenge is clear for Rockwell's long-term success. We must create a work environment where diversity is a core value … an environment in which the views and values of all are respected, and an environment in which management at all levels is committed to leveraging our diversity to our competitive advantage." ▪

> The company that started in Iowa grew to include several locations and employ people from many different backgrounds and cultures.

NYSE
Rockwell
Collins
COL
LISTED
NYSE
Rockwell
Collins
COL
LISTED
NYSE

CHAPTER 15

"The Spin," 9/11 and rebuilding

Clay Jones would later call it "serendipity" that Rockwell Collins had successfully merged its commercial and government businesses and put in place so many important initiatives by the time the company was spun off from Rockwell International.

When plans for "The Spin," as it came to be known, were announced in December 2000, the Enterprise Resource Planning system called SAP was in place throughout most of Rockwell Collins. Lean Electronics was in its third year, and considerable progress had been made in driving out waste and reducing costs.

‹ The New York Stock Exchange on July 2, 2001.

› The "Charging Bull" bronze sculpture by Arturo Di Modica near Wall Street.

A Core Process Optimization (CPO) initiative had examined and continued to refine the five core processes: strategic and financial planning, pursuit and order capture, design and development, manufacturing, and product support.

"All of that was put in place before we did the spin," Jones recalled. "We were well prepared to go into a spin because a lot of the heavy lifting had gone on under the auspices of our parent company."

"The Spin" would have taken place anyway. Jones said it was simply fortunate that business conditions had been favorable for investing in all that heavy lifting between 1996 and 2001.

Serendipity.

Rockwell Collins shares began trading on the New York Stock Exchange on July 2, 2001. After Jones rang the opening bell, the stock (COL) opened at $23.51. It closed that day at $23.66 — an auspicious gain of 15 cents.

The Collins story had come full circle.

"The Spin"

Continuing its strategic evolution, Rockwell International announced on December 8, 2000, that its board of

> Rockwell Collins shares began trading on the New York Stock Exchange on July 2, 2001. The stock (COL) opened at $23.51.

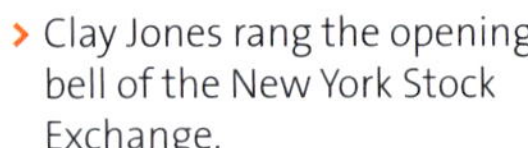

> Clay Jones rang the opening bell of the New York Stock Exchange.

directors had approved "in principle" the tax-free spinoff of its Rockwell Collins Avionics & Communications business unit to shareholders. Following the spinoff, Rockwell Collins and Rockwell Automation would both stand alone as publicly traded companies.

Don Davis, Rockwell's chairman and CEO, said, "Through separating our automation and avionics and communications businesses, we believe that each company will make decisions more quickly, deploy resources more rapidly and efficiently, and operate with greater agility. This strategic move will also directly benefit our employees, as equity-related compensation will now be directly linked to the performance of each respective business."

In a letter to Rockwell Collins employees, Jones said, "Each of us should feel very proud. It is our ability to work together and deliver world-class performance in this organization that has given Rockwell the confidence to pursue a spinoff of our business to its shareowners."

In the months leading up to the official spinoff date of June 29, 2001, there was much speculation in the media that both Rockwell Collins and Rockwell Automation would be vulnerable to being acquired by some third party. The Wall Street Journal and other newspapers, however, reported that structuring the move as a tax-free transaction for shareholders would deter acquisition attempts for two years. An acquisition in that period could have resulted in big tax bills for shareholders.

Shareowners of record at the close of business June 15, 2001, received one share of the new Rockwell Collins for each share of Rockwell International stock they held.

Preparing for the impending spinoff, Rockwell Collins announced two key leadership appointments on May 14 as part of a "sweeping organizational change." Bob Chiusano was named executive vice president and chief operating officer of Government Systems. In addition to leading government business activities, he became responsible for the company's international subsidiaries and the China Business Development office. Neal Keating was named executive vice president and chief operating officer of Commercial Systems, putting him in charge of the Air Transport, Business and Regional, and Passenger Systems businesses.

> Bob Chiusano was CACD vice president and general manager and later served as executive vice president and COO of both Government Systems and Commercial Systems.

The reorganization put operations, finance, human resources, corporate development, engineering and technology, Collins Aviation Services and the eBusiness and Lean Electronics organizations under the umbrella of Shared Services. Executives in each of those areas would report directly to Jones as president and CEO.

Announcing the organizational changes to employees, Jones acknowledged that the company had been through "tremendous change" over the previous five years, adding, "But we have emerged as a stronger company with excellent growth opportunities. The challenges and the changes we will face as an independent public company will be even more significant."

Kent Statler noted that when the Collins businesses had good years under Rockwell International, the impact was sometimes offset by lower performance in other parts of the corporation. Employees could not readily see the impact of their good work on the stock price. When employees had the opportunity to own Rockwell Collins stock, Statler said, there was a notable difference. After the spinoff, employees had a stronger sense of ownership and many of them might be found looking up the stock price. "You never saw that when we were part of Rockwell International," he said.

Rockwell Collins entered the new, independent era with more than 60 percent of its business in commercial aviation. Two-thirds of the business was domestic and about a third was international. The portfolio had resulted in 15 percent growth over the previous four years — about a third through acquisitions and two-thirds internal growth. Even in 2000, when Boeing produced 100 fewer aircraft than it had in 1999, revenues grew by 5 percent and profitability by 6 percent. Looking ahead, the markets were expected to remain strong.

In comments to analysts, Jones pointed out what were expected to be four key growth areas: next-generation avionics that would use a common core architecture applied to diverse markets; free flight based on satellite technology; "e-flight" development building on the in-flight entertainment base; and service and support, which historically had been addressed almost as "an afterthought."

"We are attempting to do something at Rockwell Collins that no one in the history of our industry has ever done," he said. "That is, to develop a common core avionics architecture that can be used and adapted with sensors that will serve all of our markets." He cited winning of the KC-135 GATM program and new positions in business and regional jets as evidence that the concept was being accepted in the market.

Rockwell Collins would stay close to that path, but before long there were new realities to take into account and strategic decisions to be made.

> The cockpit of the Bombardier Challenger 300 was the first to use the full Pro Line 21 avionics suite, which represented an important step in Rockwell Collins' common core architecture development.

> The Challenger 300 in flight.
Photo by Paul Bowen.

› Searchlights commemorated the World Trade Center towers in the "Tribute in Light."

U.S. Air Force photo/Denise Gould.

September 11, 2001

When terrorists used hijacked airliners to attack the United States on Sept. 11, 2001, Rockwell Collins was just 74 days into its new independence. The effect on the commercial airline market was immediate, with a dramatic reduction in travelers and a resulting sudden, sharp decline in orders. Aircraft manufacturers laid off thousands of employees, and Rockwell Collins stock fell to $12.95 per share within days.

While Rockwell Collins ended the year with $2.8 billion in sales — a 12 percent increase over 2000 — revenue expectations from commercial aviation were reduced going forward. Company executives responded quickly, taking measures that were later credited with minimizing the long-term impact on the business. In the fourth quarter, the company announced plans to reduce its workforce by approximately 2,800 employees, and salaries and bonuses were frozen.

The expected transition to free-flight, satellite-based navigation slowed in the wake of the September 11 attacks, as did other commercial segments of the business, including the market for in-flight entertainment and information management products. More resources worldwide were being focused on safety and security rather than on technology that would allow more direct, fuel-efficient routes.

Rockwell Collins shifted its strategic emphasis toward government business to offset the impact of the commercial downturn. Its largest product portfolio — communications, navigation and situational awareness (CNS) products and systems — had applications for both government and commercial customers. The company was well established as a leader in military electronics, including data links development through its JTIDS and MIDS participation. It continued to be the leading source of radio communications and GPS technology for both airborne and ground force navigation, and also had significant content in precision munitions.

The acquisition of K Systems, which had been completed just a week before the spinoff was announced in December 2000, increased Rockwell Collins' already substantial strength and presence in the military sector. Its San Jose-based Kaiser Aerospace and Electronics subsidiary had supplied more than 21,000 display systems for tactical aircraft, including the Comanche helicopter and F-18 fighter aircraft and the new Joint Strike Fighter. Its core products included head-down, head-up and helmet-mounted displays for pilots. Another Kaiser business, Electro-Optics, provided helmet-mounted displays for soldiers. Kaiser also controlled 50 percent of Vision Systems International, whose core business included helmet-mounted displays for several fixed-wing military aircraft.

The adjustment in emphasis, combined with some key wins in Government Systems in fiscal year 2002, made for a marked change in the balance of the business. Government business accounted for 45 percent of sales that year while 55 percent was commercial. The biggest success was Rockwell Collins' selection as part of a Boeing team to produce the industry's first Joint Tactical Radio System (JTRS)-compliant communications systems for the U.S. Army. The contract required the development of a software-reprogrammable radio for ground vehicles and rotary-wing aircraft.

In February 2002, Rockwell Collins Flight Dynamics contracted with Boeing to develop dual head-up display (HUD) systems for the U.S. Air Force C-130 Avionics

The New York Times

U.S. ATTACKED

HIJACKED JETS DESTROY TWIN TOWERS AND HIT PENTAGON IN DAY OF TERROR

A CREEPING HORROR

President Vows to Exact Punishment for 'Evil'

A Somber Bush Says Terrorism Cannot Prevail

Awaiting the Aftershocks

> The deadly terrorist attacks using hijacked airliners had an immediate impact on the commercial airline market.

> Rockwell Collins provided helmet-mounted displays for pilots in addition to head-up and head-down cockpit systems.

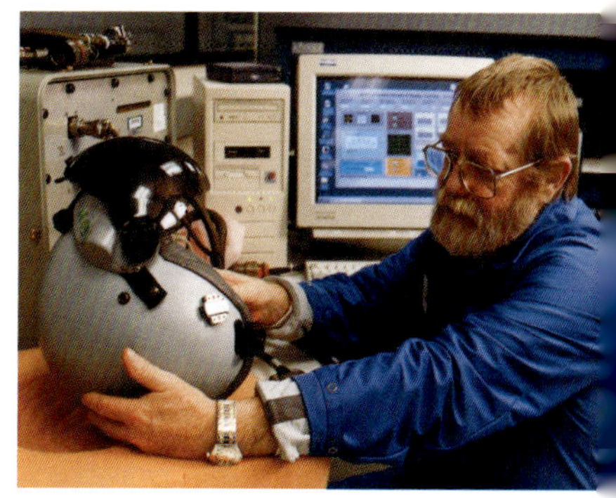

> The flight deck of the Blackhawk MH-60 helicopter (right) was upgraded in the Common Avionics Architecture System (CAAS) program using technology initially developed for the commercial avionics market.

Modernization Program (AMP). Boeing was the prime contractor to upgrade the avionics systems for more than 500 C-130 aircraft. Rockwell Collins was selected to provide other displays for the AMP, as well, and also provided HUD displays for the Air Force's C-130J and the U.S. Navy's C-40.

Rockwell Collins also was selected to provide avionics hardware for a U.S. Army Special Operations helicopter avionics upgrade program in 2002. The Common Avionics Architecture System (CAAS) program was designed to provide Special Forces helicopter fleets with an open, common avionics architecture with reusable components. Rockwell Collins' solution, as was becoming increasingly common, used technology that originated in the commercial business and adapted it to meet military requirements.

Even with the commercial slowdown, Rockwell Collins Commercial Systems generated significant new business wins in 2002, including the selection of Pro Line 21 flight deck avionics on the Cessna Citation CJ3 and the Gulfstream G150. The company also continued to acquire new capabilities that would result in still more extensive positions in the business jet market. To augment its integrated cockpit avionics systems, it entered the cabin electronics segment of the business with the acquisition of Airshow, Inc. Airshow's product line included systems for business support, information networks, passenger entertainment and cabin environmental controls — the office in the sky. Bombardier Aerospace was among Rockwell Collins' first "Airshow 21" customers when it selected the product for its Global 5000 intercontinental business jet.

The economic impact of the September 11 terrorist attacks had been most significant in North America, where the commercial air transport market was still struggling to recover late into 2002. Then, a new challenge emerged in Asia in the form of an epidemic of severe acute respiratory syndrome (SARS). Originating in China, it spread to more than two dozen other countries in Asia, Europe and North and South America before subsiding months later. In the process, it had significantly delayed recovery in commercial air travel.

> The Airshow product line included moving map and flight information displays for business aircraft.

> The success of the Pro Line 21 avionics system continued with its selection by Cessna for the Citation CJ3 (above).

In-flight photo by Paul Bowen.

Meeting other challenges

In the same time frame, changing health care and financial markets were presenting other types of challenges. Health care costs in the United States had begun skyrocketing, climbing at more than 20 percent annually, said Melodee Webb, Rockwell Collins compensation and benefits vice president. Among other changes, the company negotiated pricing on services with local providers, capped the retiree medical benefits program, and narrowed its health care provider network.

The post-9/11 layoffs, along with changes in health care and retirement programs, helped the still young, independent company maintain its financial stability and protect its customers and shareholders. Soon, the company took a new look at how to address important workforce issues.

Ron Kirchenbauer, who joined Rockwell Collins in April 2003 as senior vice president of Human Resources, said he felt the new, independent company needed a better-defined people strategy to help it move forward. It would also serve to help offset the "wave" of changes that had adversely affected the workforce. In May 2003, the Value Proposition for People was presented as part of the Human Resources strategic and financial plan. It was soon integrated into the company's strategic framework for accomplishing its objectives.

"In Human Resources, we have two primary responsibilities," Kirchenbauer said. "To help our people be all they can be, and help the business be all it can be. The Value Proposition for People will help us achieve both."

The proposition refined the employee benefit program, which became more flexible to meet people's changing needs as their careers and lives progressed. It also addressed the need to attract and retain talented people in a variety of other ways. The company's defined retirement benefit plan, which had been frozen in the face of a "precipitous drop" in funding, was replaced with an enhanced defined contribution plan that would be less subject to increasingly volatile financial markets.

> Ron Kirchenbauer implemented the company's "Value Proposition for People" to help attract and retain employees.

The Proposition

Our Value Proposition for People is our overall strategy to achieve our "talented and motivated people" goal — a key component of the Rockwell Collins Vision Roadmap.

Four broad initiatives make up our Value Proposition for People:

Flexible Benefit Choices

Provides each of us with flexibility and choice regarding our benefits decisions, while maintaining a basic level of protection for everyone.

Leadership Development

Encourages and supports the development of more and more capable leaders. The programs and tools are tailored to meet the needs of individual contributors, entry-level, mid- and executive-level leaders.

Diversity

Values and respects differences in our workplace. We strive to recruit, retain and develop talent to create an environment that fuels innovation and builds a stronger company.

Total Management

Integrates how we acquire, evaluate, reward, and develop our people. ▪

Jones said there were two things that kept him awake at night as the commercial airline market began to improve in late 2003 and early 2004. Neither of them was related to "The Spin" or 9/11 and its aftermath. Both, in fact, were based on the company's pre-spinoff performance from 1997 to 2000. In those years, he recalled, "We were chronically late in hiring people to populate the development programs we had, specifically in engineering. As a result, our performance suffered." The Lean initiative hadn't been fully implemented at that point, and core process optimization was still in progress, but being short of people was a significant impediment.

"I'd seen it happen," Jones said. "It stifled growth, and I didn't want it to happen again."

Recruiting, leadership development and other talent management efforts under the umbrella of the new Value Proposition for People delivered results. More than a third of Rockwell Collins employees were new to the company over the next six years, a period that saw the engineering force alone double from roughly 3,000 to 6,000.

A new initiative also was launched to address the other critical issue: chronic parts shortages. "We had to get our supply chain aligned to ensure that we could build the product on time and not suffer those shortages," Jones said.

By the end of 2005, Rockwell Collins had been recognized by *Purchasing* magazine with its 21st annual Medal of Professional Excellence Award for streamlining its procurement and supply chain management practices. The magazine's editor in chief said, "We have the opportunity to observe procurement operations in a wide variety of industries, and we believc Rockwell Collins is among the best of the best."

› Rockwell Collins was recognized as "among the best of the best" in 2005.

Phenomenal growth begins

In fiscal year 2002, the company's first full year as an independent company — which Jones referred to as a "trial by fire" — Rockwell Collins had total sales of just under $2.5 billion, 12 percent lower than the previous year. Commercial Systems sales were down by 21 percent, while Government Systems sales grew 4 percent. Sales overall would grow by $50 million in 2003, then to more than $2.9 billion in 2004, with earnings per share of $1.67 and a 52 percent total shareowner return.

Continued improved performance and a series of acquisitions would bring sales to nearly $4.8 billion in fiscal year 2008. In that period, Rockwell Collins invested significant resources in research and development to stay on the leading edge of technology in aviation and communications. Investments in R&D averaged nearly 19 percent of sales for fiscal years 2002 through 2008, when R&D reached $900 million.

Value Proposition for People

The Value Proposition for People was established as part of the Rockwell Collins post-spinoff strategy for achieving its goal of employing "talented and motivated people." Its framework included four broad initiatives that were integrated into the Vision Roadmap for guiding the company into the future. All four would have to work well for Rockwell Collins to succeed in the war for talent.

One initiative was focused on flexible benefit choices to attract new people and keep them on board. Others addressed leadership development, diversity, and talent management.

The Flexible Benefit Choices program was designed to allow employees to choose the mix of benefits that best met their individual and family needs. In addition to medical, dental and vision plans, a retirement savings program and life and disability insurance, Rockwell Collins used its buying power to offer access to legal services, auto and home insurance and even pet insurance.

The company also developed a comprehensive leadership roadmap, which shows employees what skills they need to have to be able to succeed at each level of responsibility. It also shows what leadership courses are required and what elective courses are available for inclusion in a leader's personal development plan. Both formal and informal mentoring programs also were started to develop leaders and help employees reach their career goals.

Diversity had been recognized earlier as something that warranted attention, but it took on greater urgency as competition for top engineering graduates heated up — and after Jones' admonition that the company could not afford to get caught short on talent as it had in the past.

The need for understanding and welcoming diversity was seen as particularly important in Cedar Rapids with its small minority community. Recruiting efforts often were successful, but new employees who came to Iowa from abroad — or even from other, more diverse areas of the United States — did not always feel welcome. Some struggled to fit in, and human resources personnel noted that a disturbing number of talented people were leaving after just two or three years as a result.

During the post-9/11 downturn, a cross-functional team was formed to address diversity issues from a formal business perspective. Rod Dooley, who helped develop the diversity "business case" and was later named vice president for talent management and diversity, said the company recognized that understanding diversity was going to be "very crucial" if it expected to meet its goals for double-digit business growth.

"We needed to make sure we had the talent to support that," he recalled. The demographic change taking place domestically was expected to have a huge impact. Dooley pointed to studies showing that about 70 percent of new entrants in the workforce through 2010 were expected to be women and people of color. In addition to changing demographics, diversity was also coming to be recognized as a key to innovation.

"Having people of different backgrounds around the table is going to drive innovation," Dooley said, because people with different experiences can address a problem in different ways.

In 2004, the company initiated a new diversity strategy as part of the Value Proposition for People to help it attract and retain employees, promote inclusion and communicate effectively on a global basis with employees and customers alike.

Several employee network groups were established to foster connections and a sense of community to help retain employees. The groups included African Americans of Rockwell Collins, Friends of Asia, the Latino Employee Network, the New Hire Network and the Women's Employee Network. Rockwell Collins also encouraged diversity among its suppliers.

The diversity strategy reached beyond the workforce and suppliers into Cedar Rapids and Iowa City and other Iowa "Technology Corridor" communities. Rockwell Collins was instrumental in the birth in 2005 of a new non-profit organization called Diversity Focus. It is sponsored by several large employers as well as smaller companies, which all have a stake in helping the community become more welcoming and accepting of diversity.

The talent management component of the Value Proposition for People integrated the ways Rockwell Collins went about acquiring new talent, evaluating their performance and rewarding them for achievement, and developing their talents and skills throughout their careers.

Employee performance evaluations took into account not only what goals were achieved, but how they were achieved to ensure that employees operated in keeping with the company's values of teamwork, innovation, integrity, customer focus and leadership. ▪

› Rockwell Collins' "Diversity Report" highlighted strides made in valuing and leveraging differences among people to build a stronger company and fuel innovation.

Vision Roadmap

The Rockwell Collins Vision Roadmap incorporates a group of five values, four goals and a strategic framework that ultimately defines Rockwell Collins, its beliefs and its future.

Vision Statement

Working together creating the most trusted source of communication and aviation electronic solutions.

Values

Teamwork:
We know that the best ideas and results are created when we work together. Therefore we embrace diversity; we support each other; and we take ownership for the performance of our team and ourselves.

Innovation:
We understand that the best source of growth is the creativity of our people. We support that creativity through investment, process efficiencies, professional development and knowledge management.

Integrity:
We will always be ethical and honest with our stakeholders and each other and never compromise the trust placed in us. This includes complying with all laws governing our corporation, our stewardship of the environment and being good citizens in the communities where we live and work.

Customer Focus:
Our reason for being is to create customer and shareowner value. We achieve this value by helping our customers be successful and always doing what we say we are going to do.

Leadership:
We take personal responsibility for making our company successful. Each of us has multiple opportunities to please a customer, reward a shareowner, help a colleague and do our job better. We must take full advantage of these opportunities.

Goals to Support Vision

- Superior customer value
- Sustainable and profitable growth
- Global leadership in served markets
- Talented and motivated people

Strategic Framework to Accomplish Goals

- Operational Excellence through Lean Electronics
- Optimize the core business
- Expansion beyond the core
- Complementary alliances and acquisitions
- Value Proposition for People

› Greg Churchill was named Government Systems executive vice president and COO in 2002.

Growth after the spinoff was fueled by continued success in development and marketing of CNS products and integrated avionics systems — including Pro Line 21 and Flight2 — major new commercial programs with both Boeing and Airbus, key Government Systems programs, and several important acquisitions. A new focus on building up the company's service and support functions also made a major contribution as its global network expanded.

Some key leadership changes were made in 2002 as the business adjusted to the new economic environment. Greg Churchill was appointed executive vice president and COO of Government Systems. Bob Chiusano, who had held that position, became executive vice president and COO of Commercial Systems. Jones became Rockwell Collins' chairman.

The "Dreamliner"

Chiusano said one of Commercial Systems' most important tasks was developing a strategy to address Boeing's impending launch of the 787 "Dreamliner" program. In December 2002, Boeing announced its intention to build a "super-efficient" mid-size, wide-body airliner using advanced technology and an unprecedented percentage of composite materials — as much as half of the aircraft's primary structure. Boeing predicted sales of more than 3,000 of the long-distance airliners over the next 20 years.

The pursuit team used an approach that started with focusing on "the customer's most important requirement," then devising a proposal to deliver what the customer wanted. The approach had been successful in the most recent Government Systems wins. Ultimately, the Rockwell Collins avionics proposal incorporated elements of the common core architecture developed for the Pro Line 21 product line in Business and Regional Systems and applied them to the 787 Dreamliner.

› Rockwell Collins' investments in head-up and head-down display technology, which it integrated for the first time in the Boeing 787 cockpit, helped win the company a commanding position on the new aircraft.

Boeing photo.

› Engineers at work in the 787 Lab at Rockwell Collins.

In 2004, Boeing announced the selection of Rockwell Collins to provide the 787's displays, communications and surveillance systems, pilot controls and core network cabinet, and to assist in development of the aircraft's common data network.

"Nobody expected us to come in and win the positions that we did," recalled Kelly Ortberg, who later succeeded Chiusano. "That was a major shift in our relationship with Boeing. Not only did we reestablish ourselves, we won more content than we'd had on any other airplane. That was a major sea change for us."

He said being awarded the fully integrated pilot controls system was particularly rewarding because Boeing had never outsourced that element before.

"It was the first time we integrated head-up and head-down displays, which played very big," he noted. The LCD position was a big surprise to many in the industry who thought Rockwell Collins had abandoned that market.

Ortberg said the keys to winning major business with Boeing had been continuing to invest in displays technology and leveraging the core capabilities developed for the business aviation and government markets. The business proved to be worth getting, and while the 787 experienced significant delays, most were related to construction of the composite fuselage and not to Rockwell Collins avionics.

"We did a good job of getting our piece of work on schedule, on budget," Ortberg said. Much of the Rockwell Collins 787 work was done in various Iowa facilities as well as in Melbourne, Irvine and Portland.

Rockwell Collins and the Boeing 787

A 2009 news release described the extent of the company's position on the 787:

Rockwell Collins content on Boeing's 787 airplane includes an integrated display system featuring five 15.1-inch diagonal LCD displays — four across the flight deck and one in the control stand for emulation of the Control Display Units (CDU) — as well as dual LCD head-up displays (HUD). The system utilizes cursor control devices and a multi-function key pad for data entry and retrieval. Rockwell Collins employed advanced model based development processes and tools to make significant reductions in development cost, cycle time and life cycle costs typically associated with display systems.

The 787 flight deck also features Rockwell Collins' latest generation of pilot controls — further advancing the company's leadership in design, integration and manufacturing of this technology. Rockwell Collins has developed the control stand including auto throttles, and pitch, roll, yaw and primary flight controls, as well as their interfaces to the aircraft's fly-by-wire systems. The modular design of the pilot controls will simplify installation and maintenance. This new system meets Boeing's objective of providing operators with a look and feel similar to the Boeing 777, while achieving significant weight savings.

Rockwell Collins provides a newly developed Integrated Surveillance System (ISS) for the 787. This highly integrated system includes functions such as hazard detection, traffic alert and collision avoidance, Mode S surveillance, and terrain awareness and warning capabilities.

The 787's communication system includes Rockwell Collins' VHF-2100, SAT-2100 and HFS 900D. The lighter weight, highly reliable VHF-2100 is VDL Mode 2 capable with future growth to VDL Mode 3 and 4. The new, smaller and more reliable SAT-2100 supports the International Civil Aviation Organization's safety services, as well as three channels of voice communications and offers growth to future Inmarsat Swift high-speed data capabilities. As part of the communications package, Rockwell Collins is also providing a state of the art digital flight deck audio system, and the cockpit voice and flight data recording system.

› An early rendering of the Boeing 787 flight deck.

The Core Network, offered as basic on the 787, leverages Rockwell Collins' investment in Information Management products. This next generation of the Core Network plays a key role in Boeing's objective to 'e-enable' the entire aircraft. Utilizing commercial open standards, the Core Network hosts a wide range of third-party applications, and manages onboard information flow, to improve airline operational efficiency.

The 787's Common Data Network (CDN) advances Rockwell Collins' leadership as a supplier of advanced networking technologies. As a key component of GE Aviation Systems' Common Core System, the CDN is a high integrity, bi-directional fiber optic and copper network that uses ARINC 664 protocols and standards to manage the information flow between the aircraft's onboard systems. Based on commercial Ethernet technology, adapted to the avionics environment, the integrity and deterministic characteristics of Rockwell Collins' CDN allows systems integrators to utilize this network for systems requiring a high level of data criticality. The CDN offers significant improvements over current generation data buses including expanded connectivity, higher data rates and significant reductions in aircraft weight when compared with point to point topologies. ▪

> More Rockwell Collins systems were selected for the new A350 XWB than for any earlier Airbus platform.

Airbus S.A.S. photo.

Airbus position takes off

Rockwell Collins' impressive win with the Boeing 787 got the attention of Airbus executives, among others, leading to what Ortberg called "a huge market lift."

Airbus previously had made individual equipment and parts purchases, and Rockwell Collins equipment was already in use on the A320, A330, A340 and A380. Then, in November 2007, Rockwell Collins announced that it would provide the trimmable horizontal stabilizer actuator (THSA) for the new, extra-wide-body A350 XWB, which Airbus planned to enter into service in 2013. A subsequent award, announced in April 2008, included three significant avionics packages: the fully integrated Communication Global Network package, Avionics Data Network, and landing systems. At the end of July 2008 came the announcement that Airbus had selected the company to provide information management and navigation capabilities, as well.

> The trimmable horizontal stabilizer actuator (THSA) that Airbus chose for the A350 XWB.

All told, the A350 XWB would have more Rockwell Collins content than any previous Airbus platform. The work would be completed in a variety of locations, including Cedar Rapids; Irvine, Texas; Melbourne, Florida; and Toulouse, France. The potential value of the program was expected to be as much as $2.5 billion. Another important distinction for Rockwell Collins was its selection as standard equipment.

"For the first time, the systems we provide for Airbus will not be airline-selectable," said Jeff Standerski, vice president and general manager for Air Transport Systems. "Each time a new airline places an order for an Airbus A350 XWB, the aircraft is guaranteed to be equipped with all of the Rockwell Collins systems we have won."

Success in regional and business aircraft

Rockwell Collins also established itself firmly in the growing regional jet market, most notably with the selection in 2003 of the Pro Line 21 avionics system by AVIC I Commercial Aircraft Company of China, which was building the country's first homegrown commercial aircraft. The ARJ21 flight deck would include five Collins high-resolution liquid crystal adaptive flight displays, integrated processing and Ethernet network communications.

Denny Helgeson, who was vice president and general manager of Business and Regional Systems, noted, "It is fitting that the ARJ21 (Advanced Regional Jet for the 21st century) be equipped with the avionics system for the 21st century."

The Pro Line family of avionics, which originated in 1970 under the "Low Profile" name, evolved into Pro Line Fusion™, which was introduced in November 2007. Within months, it had been selected for use on a broad range of business and regional aircraft, including Bombardier's Global Express XRS, Global 5000, C-Series and

› The ARJ21 flight deck with Pro Line 21 avionics.

› Bombardier's Global Express XRS was one of the first aircraft to use the Pro Line Fusion system (flight deck below), which Rockwell Collins introduced in 2007.

In-flight photo by Paul Bowen.

› The Embraer Legacy 500 (top, and flight deck below) was one of many business and regional aircraft equipped with the Pro Line Fusion system's large, high-resolution LCDs and MultiScan weather radar.

Lear 85; the Embraer Legacy 450 and 500; the Gulfstream G250 and Mitsubishi Heavy Industries' new regional jet, the MRJ.

The new Pro Line Fusion system included high-resolution, 15-inch LCD displays, MultiScan™ weather radar, and more intuitive, point-and-click controls, and context-sensitive information windows in addition to integrated cabin and flight deck avionics diagnostics.

In the seven years following "The Spin," Rockwell Collins Business and Regional Systems won a remarkable 21 of 22 bids in competitions for new aircraft avionics.

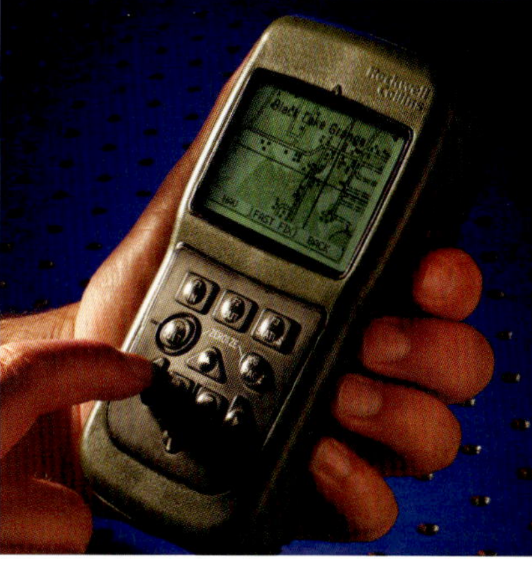

> The Defense Advanced Global Positioning Receiver (DAGR) included many improvements over the PLGR, its predecessor.

Government market gains

Winning a role in the JTRS program in 2002 marked yet another significant change in Rockwell Collins, positioning it as a provider not only in airborne communications, but in ground communication, as well. Clay Jones said the shift in technology was "as big as the shift from analog to digital." It was an entry into an enormous potential market, that Rockwell Collins had not served previously.

The company played a prominent role in developing the JTRS architecture as part of the Modular Software Radio Consortium and developed an early JTRS prototype. Its initial contract for work on ground vehicles and helicopter communications ultimately led to a major, ongoing role and several additional contracts.

Rockwell Collins' performance in the earlier PLGR handheld GPS program, in which it delivered more than 250,000 units to customers around the world, led to its development and production of the PLGR's successor — the Defense Advanced Global Positioning Receiver (DAGR). The DAGR was smaller, lighter and more powerful, and designed to be "backwards compatible" with the PLGR receivers that had been deployed extensively in wars in Iraq and Afghanistan and elsewhere. The DAGR receiver was capable of using up to 12 channels to lock in signals from available GPS satellites, giving soldiers and other users more reliable position information in challenging conditions and terrain.

> Medallions commemorated production of the 100,000th DAGR unit. Later, sales more than tripled.

Weighing less than a pound, with batteries, it included anti-jamming improvements, mapping, a new graphical user interface, and the Rockwell Collins Selective Availability Anti-Spoofing Module (SAASM) to protect the military GPS data from enemy interference. The full-rate DAGR production contract was awarded in October 2003. The company later introduced a non-military version of the DAGR for use in commercial markets.

In 2009, Rockwell Collins delivered its 300,000th DAGR receiver, a landmark that was celebrated with a special ceremony at the production facilities in Coralville. The company had expected to produce no more than 100,000 units, but the ongoing wars kept demand high. As Col. Dorothy Taneyhill, U.S. Army project manager for navigation systems, said, "Continued delivery of the DAGR to our deployed forces is vital for accurate, expedient situational awareness, and to achieve mission effectiveness on the battlefield."

The company also received an additional $450 million follow-on contract in 2009 to provide DAGR receivers for the U.S. Air Force through 2016.

Other major military contracts for Government Systems included awards for work on the Integrated Computer System (ICS) for the U.S. Army and its Future Combat Systems (FCS) program. The ICS is the common computing environment for most FCS platforms. FCS consists of a network of sensors, UAVs and both manned and unmanned ground vehicles and equipment.

Rockwell Collins teamed with General Dynamics Advanced Information Systems to win a $153.9 million contract for accelerated ICS development to get the system in place on Bradley fighting vehicles, Abrams main battle tanks, and Stryker and Humvee vehicles.

> Rockwell Collins marked delivery of the 300,000th DAGR receiver in 2009 with a special ceremony in Coralville and remarks by Col. Dorothy Taneyhill (left). Deb Hansen, IBEW Local 1634 president, accepted a commemorative plaque on behalf of the DAGR production team.

› Barry M. Abzug, Ph.D., was named senior vice president, Corporate Development, in 2001.

› With the acquisition of Evans & Sutherland's simulation imaging business and NLX's full-scale simulator capabilities, the stage was set for Rockwell Collins to enter the markets for both military and commercial aircraft training.

Growth by acquisition

The acquisitions in displays technology, in-flight entertainment and cabin systems that preceded and closely followed "The Spin" were just a start. Several more would be made over the next few years to meet various strategic needs.

Barry Abzug, Ph.D., senior vice president in charge of corporate development since 2001, described the Rockwell Collins approach to acquisitions as "disciplined." Each acquisition was either in the company's core business or in adjacencies that closely supported strategic goals.

Three of those acquisitions helped Rockwell Collins build a significant role as a provider of flight-simulation technology and training expertise, an important addition to the company's services offerings: NLX in 2003, Evans & Sutherland's simulation business in 2006, and SEOS in 2008. The resulting revenues would make Rockwell Collins Services — the unit that came to include the simulation and training business — a major factor in the company's overall success. From less than $200 million in business in 1998, RCS would grow to nearly $1 billion by 2009.

NLX was a training system provider with expertise in the analysis, design, manufacture, production, deployment and support of advanced operator and maintenance training systems. The original mission for NLX was to provide upgrade and modification services to various military training programs. The company's first contract was in the support of the U.S. Air Force undergraduate pilot training systems. Rockwell Collins Simulation and Training Solutions (STS) subsequently increased its capabilities toward full-scale training system development.

Rockwell Collins acquired Evans & Sutherland's military and commercial simulation business, which was noted for being the world leader in image generation and database development. The acquisition of SEOS, a British company, in November 2008 further enhanced the STS organization's visual display capabilities.

Other post-spinoff acquisitions addressed additional strategic priorities.

The purchase of Communication Solutions, Inc., in March 2002 expanded Rockwell Collins' product portfolio in the areas of surveillance solutions and signals intelligence (SIGINT), which is used in defense and security-related applications. Communication Solutions, which was based in Maryland, was merged in November 2006 into the company's Command, Control, Communications and Intelligence (C3I) Solutions business in Richardson, Texas.

To gain a significant strategic presence in the European military market, Rockwell Collins in April 2005 acquired TELDIX GmbH from Northrop Grumman Corporation. Based in Heidelberg, Germany, TELDIX had developed a broad portfolio of complex military aircraft computer products. TELDIX had leading positions on major European programs such as the Eurofighter Typhoon, the multi-role combat fighter Tornado, and the NH90 and Tiger helicopters.

› Electronic warfare operators use the E-2C Hawkeye simulator for training in the use of computerized sensors to provide early warning, threat analysis and control of counteraction against air and surface targets.

› With the acquisition of Athena Technologies in 2008, Rockwell Collins entered the market for UAV avionics. Athena had developed innovative flight controls and navigation systems for the unmanned aerial vehicles.

Two acquisitions in September 2006 strengthened Rockwell Collins' expertise in the increasingly important areas of wireless networking for military customers. One was Anzus, Inc., a privately owned software developer founded in 1986 and headquartered in Poway, California. It was a leading provider of real-time data exchange forwarding, display and situational awareness software for the U.S. military and allied countries. Anzus specialized in engineering, training and technical services in the areas of command and control, communications, computers, intelligence, surveillance and reconnaissance. At the same time, Rockwell Collins also acquired IP Unwired, a privately owned Canadian developer of high-data-rate HF/VHF/UHF modem and networking products and services.

The acquisition of Information Technology and Applications Corporation (ITAC), in August 2007, gave the company additional capabilities in electronic intelligence, surveillance, reconnaissance and communications. Its major customers included the Department of Defense and various national intelligence agencies.

In April 2008, Rockwell Collins entered the market for unmanned aerial vehicle (UAV) avionics with the acquisition of Athena Technologies. Its expertise was in flight controls and navigation for UAVs such as the U.S. Air Force's Predator and Reaper reconnaissance and attack drones. Its innovative flight controls technology produced a lightweight, powerful system that was expected to prove useful in commercial as well as military applications.

Later, in June 2009, the company completed its acquisition of DataPath, Inc., which had established a strong position in satellite-based communication networks by making commercial bandwidth available to the military in wartime.

› The DataPath ET 3000 Portable satellite terminal in service in Northern Iraq.

› Rockwell Collins celebrated the ongoing success of the ARC-210 radio in 2008.

Striving for success

The new Rockwell Collins era of independence brought with it numerous notable achievements in engineering, product sales and improved performance, along with honors and recognition by business and industry media and a list of product milestones. In 2003, for example, Rockwell Collins surpassed more than $1 billion in sales of ARC-210 radios.

The following year, *Forbes* magazine named Rockwell Collins to its list of "best managed" companies and declared it "the best" in the aerospace/defense industry, making special note of its emphasis on developing leadership skills. The honor was repeated in 2006. *Aviation Week* named Rockwell Collins a "top performing" company in 2007. *Business Week* did the same in both 2007 and 2008.

The company also was making significant progress in defense communications, information management, open systems architecture, and service and support.

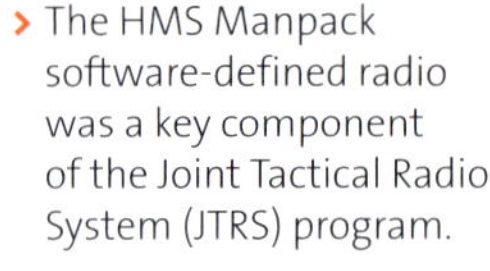

> The HMS Manpack software-defined radio was a key component of the Joint Tactical Radio System (JTRS) program.

Following its initial JTRS contract in 2002, Rockwell Collins had won additional contracts for more elements of the communication program, for "manpack," handheld and other applications as well as design work for airborne, maritime and fixed-site radios.

As travelers sought to stay in touch and informed while in the air, Rockwell Collins won important information management positions with Airbus and Boeing to provide onboard equipment that would enable access to live TV, email and the Internet. To provide similar broadband connectivity for business aircraft, the company also launched its new eXchange product. Dassault selected Rockwell Collins to provide cabin electronics for its Falcon business aircraft, and the company won expanded positions with business jets manufactured by Bombardier, Gulfstream, Raytheon and others.

A significant agreement in 2004 to provide maintenance support for Mesa Airlines' fleet of regional jets represented a "milestone" in the company's effort to grow through expanded service and support functions.

The company's sales reached $3.5 billion in 2005, a year when Rockwell Collins delivered more than 40,000 DAGR handheld GPS receivers for use by troops in Iraq and Afghanistan. They were being manufactured at a rate of 300 per day.

At the same time, the company was delivering new avionics for numerous business jets, and development work on systems for the Boeing 787 Dreamliner was "in high gear." The new Airbus 380 made its first public flights at the 2005 Paris Air Show using an open system standard Ethernet backbone from Rockwell Collins.

Using the same open systems architecture, the company also delivered its 100th KC-135 tanker avionics upgrade. The KC-135 program was touted by the Air Force as a model program for meeting or exceeding delivery and cost requirements, leading to additional upgrade programs. Those included programs for Air Force and international C-130 transports, U.S. Army and Special Operations helicopters and U.S. Navy E6-B aircraft.

> Rockwell Collins open systems architecture was used in upgrading avionics for the U.S. Navy's E6-B aircraft (below), and others.

In August 2005, the company received the industry's first Technical Standard Order approval for a Multi-mode Receiver (MMR) with Local Area Augmentation System (LAAS) functionality. The Collins GLU-925 MMR was the most advanced system to date, and the first to include LAAS and a GPS landing system (GLS). It extended the use of satellite navigation from en route and terminal operations into the arena of precision approaches. Rockwell Collins' family of MMR integrated navigation units provided the aircraft's primary position, velocity and time reference. The company fielded a variety of MMRs that included ILS, VHF Omnidirectional Range (VOR), MLS, Marker Beacon and GPS, to meet the operating requirements of the military and commercial air transport community.

Support for ground-based military forces also played an important role in Rockwell Collins' post-spinoff success. In October 2005,

General Dynamics C4 Systems chose the company to provide Helmet Mounted Displays (HMDs) for the U.S. Army's Mounted Warrior (MW) program. The MW program, which required the use of a helmet-mounted display for hands-free viewing and increased situational awareness, equipped Army crew members who were assigned to Stryker vehicles. General Dynamics selected the company's ProView™ S035 monocular for the program. Work on the helmet was done in Carlsbad, California.

> Helmet-mounted displays (HMDs) like the Pro View SO35 gave soldiers access to GPS information, maps and other data.

Clay Jones attributed the company's achievements in those and other programs, in part, to the continued, ongoing emphasis on Lean Electronics.

"It did significantly improve our program performance, to the point that now we are highly confident we can do what we say we're going to do," he said. "And it has without question changed the culture of this company and allowed us to work together as a single unit by using that driving influence to shape the structure of our company, the way we work, and the way we think about that work."

One of many Lean Electronics initiatives was the implementation in 2004 and 2005 of Life Cycle Value Stream Management to reduce design and development cycle times and maximize the value of products and services delivered to customers. The "value stream" defined a set of activities required to deliver a product or service to the customer. Engineers, Lean Specialists and others used graphical "value stream maps" to chart those activities and identify where they could be improved.

Another key Lean program, started in November 2005, was the Open Innovation Initiative to encourage the use of global technology, innovative sourcing and intelligence networks to reduce cycle times and get products and improvements to customers faster. Nan Mattai, senior vice president of Rockwell Collins Engineering & Technology, said the open innovation philosophy was meant to break the company loose from a tendency to "reinvent the wheel" on occasion.

"Open innovation recognizes that while there are 8,000 engineers here, there are millions of others innovating out there," she said. "We needed to be open to bringing those innovations in when they make sense and adapting them to provide solutions for customers. We have to be willing to collaborate inside and outside of our own four walls in order to bring the best solutions to our customers."

Mattai pointed to the Advanced Technology Center as an incubator of such ideas, expected to transition into products within three to five years, in a disciplined process to keep innovation in sync with business development and customer needs. Engineers in that center would work closely with DARPA, military research labs and the National Security Agency on numerous programs over the next several years. The company also established a new relationship with Sandia National Laboratories to transition technology developed there into advanced radar remote sensing capabilities for the military.

> Nan Mattai (right) facilitated changes that made it easier to move talented engineers between commercial and government programs in response to changing needs.

Evolution of the trademark

› 1933-1934:

› 1934-1937:

› 1934-1937: Including corporate text.

COLLINS RADIO COMPANY
Designers and Manufacturers of
Transmitters, Transformers and Speech Equipment
CEDAR RAPIDS IOWA, U. S. A.

› 1937: Experimental artwork that would evolve into the familiar bar logo of 1938. Used for a Jan.-Feb. 1937, *Communications* advertisement.

› 1938-1960:

› 1938-1960: Abbreviated, perspective and script versions used for manuals and advertisements.

› 1961-1972:

› 1973-1990:

› 1991-2006:

› 2006-present:

Building trust, building business

Rockwell Collins
Building trust every day

› With a new logo adopted in 2006 came a new brand promise.

In 2006, Rockwell Collins continued to deliver on its newly adopted brand promise of "Building trust every day" by meeting customers' needs. It made its first "red label" preproduction delivery deadlines in the Boeing 787 development program and met several key milestones for Airbus, as well. It also delivered the 50,000th DAGR to the U.S. Army to meet soldiers' needs for support in Iraq and Afghanistan.

For international military customers, the company introduced its new FlexNet™ family of software-programmable radios in a joint venture with Thales, a European defense electronics company. FlexNet radios included the FlexNet-One single-channel radio and the FlexNet-Four multi-channel radio.

To get network-centric technology to the military more quickly and efficiently, Rockwell Collins opened a new lab in Richardson, Texas, in November 2006. The Performance and Architectural Collaboration Environment (PACE) Laboratory was designed to use advanced simulations to prove military communication, sensor and navigation system concepts. PACE engineers simulated operational scenarios to validate potential solutions in the laboratory rather than in the field, reducing costs and risks.

The following year, Rockwell Collins presented the U.S. Air Force with the 200th KC-135 tanker aircraft equipped with GATM technology. It further extended its position in upgrading avionics for military tankers and transport aircraft by winning several international C-130 upgrade programs, including new business agreements in Singapore and Thailand. That brought to nine the number of countries that had selected the company's open architecture avionics for their C-130 fleets.

In September 2007, Rockwell Collins introduced its RTA-4100 MultiScan™ weather radar for new aircraft equipped with Pro Line 21 and Pro Line Fusion avionics. It was the latest product in the innovative MultiScan line and provided a 300-nautical-mile "clutter free" weather display and enhanced threat detection capabilities.

Also in 2007, the Common Avionics Architecture System (CAAS) being deployed on U.S. Army helicopters helped establish wins internationally, including the AgustaWestland AW149 and the Eurocopter German CH-53 heavy lift helicopter. The already aggressive GPS production program ramped up further in 2007, when the company delivered some 125,000 handheld and embedded GPS systems overall.

› The PACE Laboratory opened in Richardson, Texas, in 2006.

› The FlexNet-One software-defined radio, developed in a joint venture with Thales.

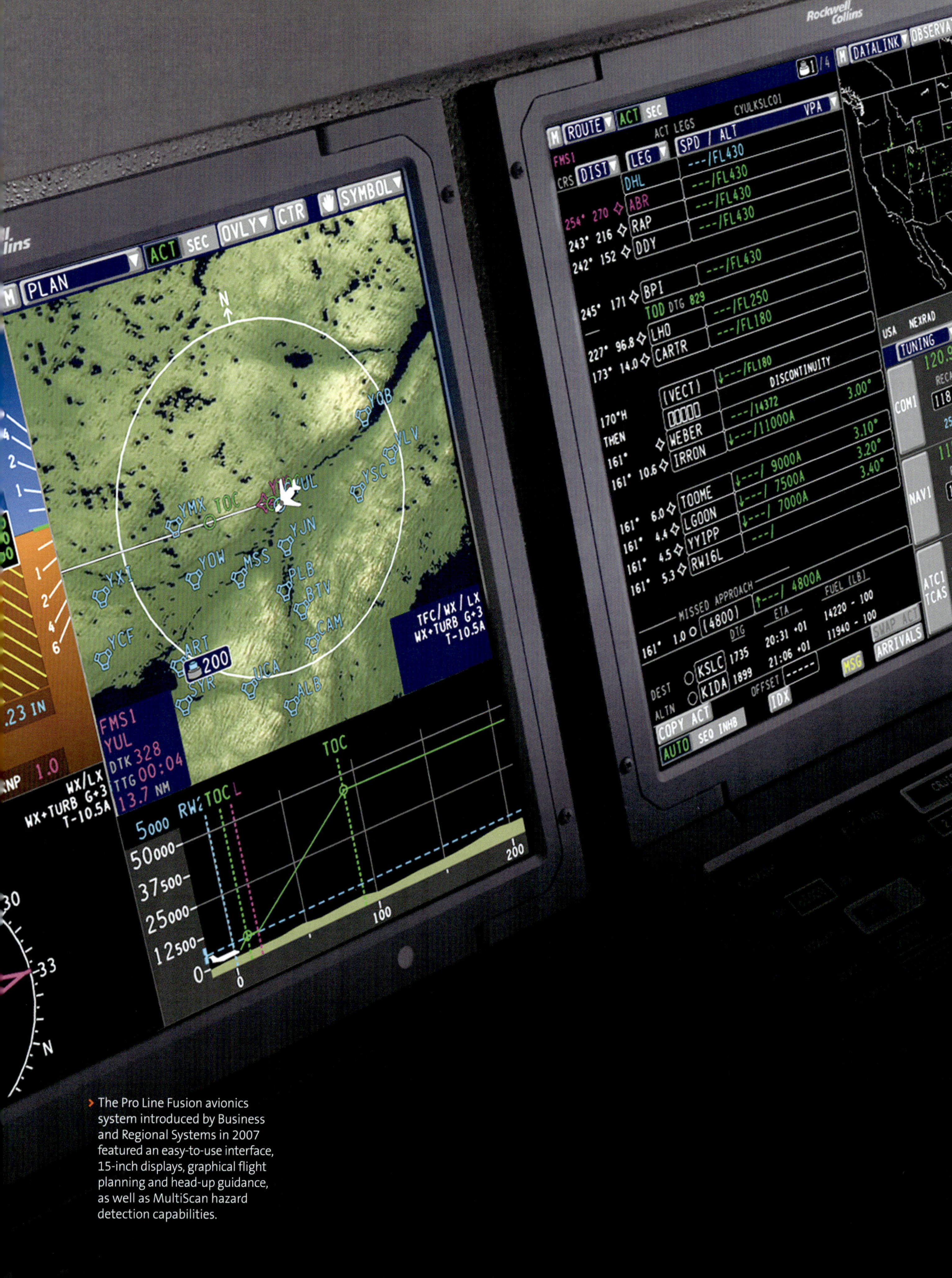

The Pro Line Fusion avionics system introduced by Business and Regional Systems in 2007 featured an easy-to-use interface, 15-inch displays, graphical flight planning and head-up guidance, as well as MultiScan hazard detection capabilities.

› Rockwell Collins leaders joined government and industry colleagues in August 2008 for the grand opening of a new company facility in Huntsville, Alabama. Pictured (from left) are: Evans Quinlivan, Huntsville Chamber of Commerce; Mike Myers, vice president of Business Development, Government Systems; Huntsville Mayor Loretta Spencer; Scott Jacobsen, manager, Huntsville Operations; Alabama Lt. Gov. Jim Folsom, Jr.; Kent Statler, executive vice president, Rockwell Collins Services; and Ken Schreder, vice president and general manager, Rockwell Collins Simulation & Training Solutions.

In the improving U.S. commercial air transport market, Continental Airlines selected Rockwell Collins avionics for more than 100 new Boeing 737 and 777 aircraft it had on order. While the new Pro Line Fusion system for business and regional aircraft was just being introduced, additional aircraft were certified with the still popular Pro Line 21 avionics suite.

The company reported "excellent" operating performance again in 2008, when sales reached $4.77 billion and its world-wide workforce numbered some 20,000 people. As a growing technology company with an increasingly global customer base, Rockwell Collins found itself looking abroad for engineering talent as well as for additional business.

The company had previously established a presence in India through a relationship with HCL Technologies Ltd. In that arrangement, approximately 300 HCL engineers, professionals and experts were working with Rockwell Collins to perform design, development and testing-related work for hardware and software-based products at design centers in Chennai and Bangalore. In February 2008, the company took the additional step of opening the India Design Center in Hyderabad to meet its growing demand for engineering talent. It planned to hire 500 people there over the next five years. The new facility's initial focus was on the design of displays and flight management systems.

Nan Mattai said establishment of the India Design Center was, in part, a response to a shortage of new engineering graduates in the United States, a case of exploring new markets for talent. The estimated 7.5 percent of U.S. college students pursuing engineering degrees was not expected to be enough.

"You go fish where the fish are," she said.

That was expected to become an even greater concern as growing numbers of engineers from the Baby Boomer generation were nearing retirement. Rockwell Collins and other players in the industry were all facing the prospect of losing what Mattai called the "deep domain knowledge" amassed during those careers.

In addition to hiring in India — and looking to China, as well, as an additional source of engineering talent — the company was developing new knowledge-management, leadership and skills-inventory programs to capture and leverage its vast inventory of expertise. It also further enlarged its U.S. footprint by opening a new 30,000-square-foot

› The India Design Center opened in 2008 as Rockwell Collins sought to meet increasing demands for engineering talent.

New spaces in the C Avenue facilities encourage collaboration among employees.

Photo by Main Street Studio.

The U.S. Green Building Council recognized new Rockwell Collins buildings on C Avenue in Cedar Rapids with its Leadership in Energy and Environmental Design (LEED) Gold Certification. The buildings were designed to minimize energy consumption and made extensive use of recycled materials. The air-handling system (below) reuses heat generated in the building's labs.

Photos by Main Street Studio.

operations center in Huntsville, Alabama, to expand its engineering, simulation and training, and product testing and support capabilities.

The company also continued its role as a leader in workplace diversity efforts. Rockwell Collins was recognized in both 2008 and 2009 as one of *DiversityInc's* "Top 25 Noteworthy Companies for Diversity." In 2008, the company received the Corporate Diversity Innovation Award during Virtcom Consulting's World Diversity Leadership Summit. It also was named a "2009 Best Diversity Company" by readers of *Diversity/Careers in Engineering & Information Technology* magazine.

Involved in communities

Rockwell Collins employees and retirees are well known for their civic involvement. The corporation also supports its various communities in a variety of ways. In 2004, it started a Green Communities program, which awarded more than 130 grants in the next four years for projects around the world. Grants ranged from $500 to $2,500 to support projects in which employees were involved.

Concern for the environment was demonstrated a number of ways. Two new buildings constructed on C Avenue were designed to meet Leadership in Energy and Environmental Design (LEED) standards to minimize energy consumption. They use "heat harvesting" technology and daylight scheduling to maximize energy efficiency. More than 99 percent of the waste generated during construction was recycled, and the buildings' contents

Left: The annual FIRST LEGO League design challenges and other employee volunteer programs benefit students in Rockwell Collins communities.

Above: Clay Jones tells a gathering about the company's Engineering Experiences program, which is designed to engage and motivate students. The varied activities support classroom learning through hands-on, interactive projects and by establishing mentoring relationships for students with employees and retirees.

consist of nearly 70 percent recycled materials. Recycling is practiced throughout the organization to reduce the company's impact on the environment.

Schools and non-profit organizations in many Rockwell Collins communities benefit from the work of the Rockwell Educational Access to Computer Technology (REACT) Program, in which thousands of computers, monitors and printers are refurbished for their use.

"We have an obligation as a corporate citizen in the communities where we operate, not only to protect, but where possible, to enhance the natural environment," said Tom Gentner, director of environment, safety and health.

Many individual employees and retirees participate in volunteer activities. More than 1,200 participate in the Engineering Experiences program, which includes the annual FIRST LEGO League design challenges. Employees volunteer more than 1,200 hours a month to community schools, much of it on company time. Nearly 200 retirees in the Cedar Rapids area are active in the Rockwell Collins Retiree Volunteers (RCRV), which was established in 1999 at the urging of a group of former employees spearheaded by Jack Hotchkiss. Mike Wilson, the group's former chairman, noted that RCRV includes many individuals who wanted to be involved more in community programs during their careers, but time and travel commitments often made it difficult.

Employees and retirees also volunteered many thousands of hours in the wake of catastrophic flooding in Iowa in June 2008. Some of their efforts to help flood victims continued on well into 2009.

When flooding struck Eastern Iowa communities in 2008, employees and retirees spent thousands of hours on recovery efforts.

Some Rockwell Collins displays used light-emitting diodes (OLEDs), which showed tremendous potential for use in new applications. Thin, lightweight and impact-resistant, OLEDs can be rolled up or folded and offer the advantages of low cost and low power requirements.

> Unmanned aerial systems, visual displays, information management and communications solutions were among Rockwell Collins' priority areas for continued investment.

Looking ahead

Along with the rest of the world, Rockwell Collins closed 2008 and started 2009 in the midst of a recession that was considered the worst in decades in the United States. Air travel had declined. A Boeing strike and repeated delays in its 787 Dreamliner program further hampered the commercial airline avionics market. The business aircraft market suffered a significant downturn following public indignation over the use of corporate jets by some companies that had received billions of dollars in federal bailouts.

Rockwell Collins' stock dipped to $27.67 early in the year before climbing back to the mid- to upper-$50s range. In aligning its manufacturing capacity with anticipated market needs, the company announced that it would close its San Jose, California, operations. Globally, eight percent of the Rockwell Collins workforce was laid off.

In July 2009, Clay Jones said the downturn was "at least the equivalent of what we went through after 9/11" and he expected the recovery to take longer. In response to difficult market conditions, the company was "tweaking" its strategies but was being well served by the balance it had established over the previous dozen years. Jones said the plan was to "stay focused on our customers and what their needs are and how we provide solutions to their needs." It would also be important, he said, to keep investing in "the right things."

Jones said the company would continue to pursue a leadership role in information management through networked communications for commercial aircraft and military operations. While building on recent acquisitions in simulation, training and visual display systems, it also

> New products included the company's iPod docking station, added to its Venue cabin management system.

‹ The Shadow 200 tactical unmanned aerial vehicle (UAV), with Rockwell Collins' Athena flight controls, served in Operation Iraqi Freedom.

would continue to invest in unmanned aerial systems and communication and computational solutions for ground vehicle and soldier applications.

Also looking forward, John Borghese, vice president in charge of the Advanced Technology Center, said some "big radical things" would be transitioning into new product applications over the next several years. Their common objectives were making communications and aviation electronics technology much, much smaller and lighter.

One was a DARPA project in which Rockwell Collins was leading an effort to miniaturize radio technology for the Analog Spectral Processors (ASP) program using micro electro-mechanical systems (MEMS). One of the first applications would be the military JTRS program.

The company also was working on a DARPA program to develop a chip-scale spectrum analyzer, and with DARPA and Georgia Tech to replicate and bond carbon nanotubes to electronic components to dissipate heat more efficiently.

Nan Mattai said other likely technological challenges ahead would lie in increasing unmanned aerial system (UAS) capabilities, sense-and-avoid systems, advances in "synthetic vision" to allow pilots to see clearly regardless of the weather or time of day, and developments such as cognitive radios that can sense and adapt to changing environments.

Rockwell Collins was making roughly a third of its sales outside the United States by 2009 and working hard to tap even more deeply in the future into fast-growing markets in Europe, the Middle East, Africa and the Asia-Pacific region. The importance the company placed on a strong global presence had been signaled early in 2008 with the establishment of the International Business and Washington Operations organization led by Walter "Woody" S. Hogle, Jr., as senior vice president.

› Rockwell Collins planned to continue development of ground vehicle and soldier applications, including systems that enabled combat units to identify positions of friends and foes through "Blue Force" tracking and data displays.

› A Head-up Guidance System composite image shows a pilot's view of the runway approach in Innsbruck, Austria, using synthetic vision technology.

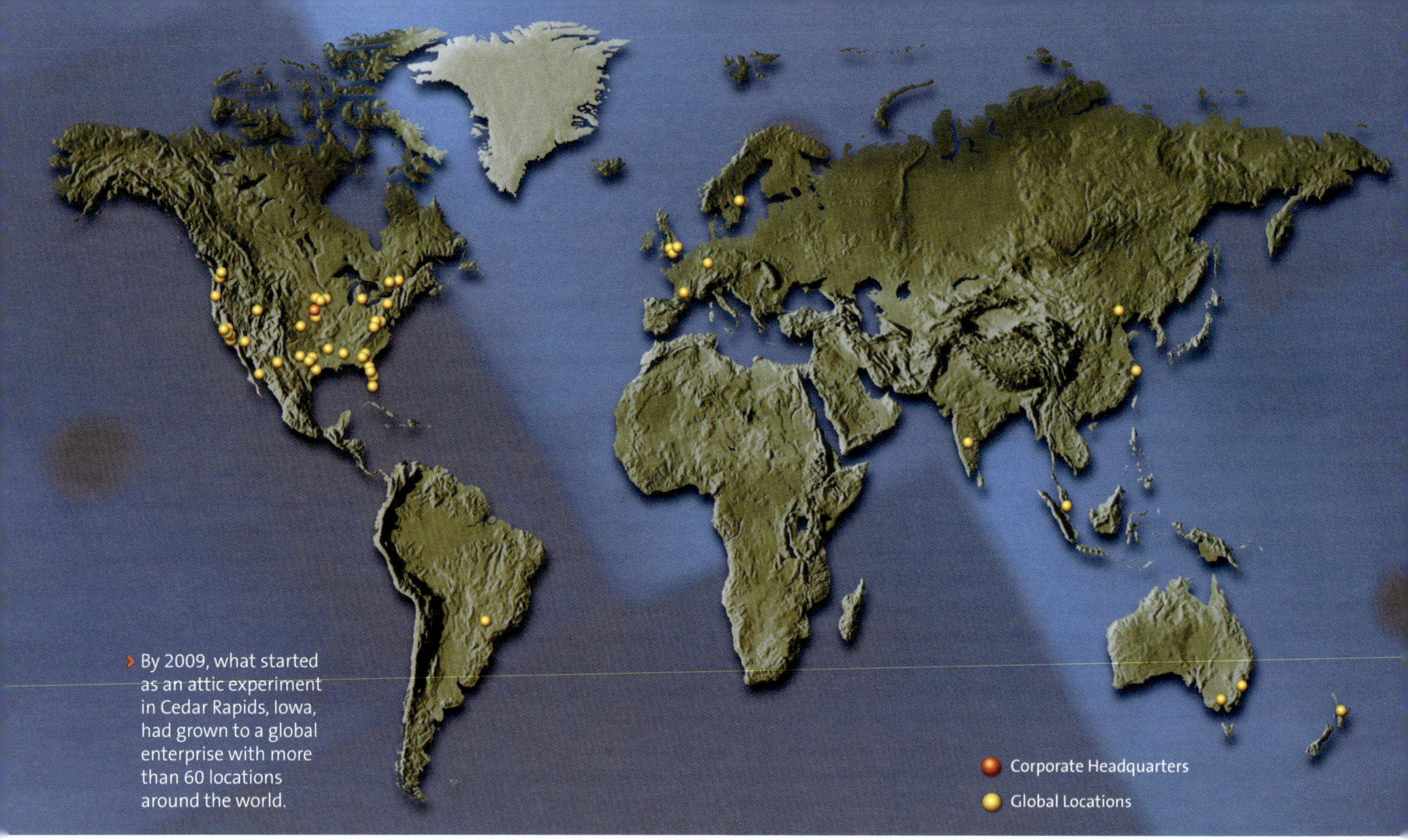

By 2009, what started as an attic experiment in Cedar Rapids, Iowa, had grown to a global enterprise with more than 60 locations around the world.

Corporate Headquarters
Global Locations

"As we grow globally," Hogle said, "more customers are going to demand that we have employees who speak their languages, understand their cultures, and live in their time zones. We need a presence in their regions to be competitive."

Jones said continued investment would be key to maintaining the company's existing market leadership and expanding into new markets.

"The strategies on which we founded this company will serve us well as we face the risks and uncertainties that remain in our markets, and they will help us deliver long-term value for our shareholders," he said.

Under the leadership of Clay Jones as chairman, president and CEO, Rockwell Collins went through a corporate transformation and experienced tremendous growth in both U.S. and international markets.

The final exam

A few years earlier, when the company was steadily winning new market share and paying out great bonuses, Jones said, he'd been concerned about people becoming complacent.

"I'm not worried about complacency now," he said in 2009. "There's nobody in this company that's complacent right now. They're concerned about their jobs. They're concerned about the economy. They're concerned about the performance of the company. They're concerned about the stock price. Keeping them on their game is not a problem."

The downturn in 2009 presented an opportunity, he said: a chance to distance Rockwell Collins from its competition — the "ultimate goal" of creating a degree of separation from the industry.

"I feel very good about what we've achieved," he said. "At the same time, I'm absolutely convinced this company has not achieved the full potential I think it's capable of, so there is a lot of work yet to do."

Turning to education for a metaphor, he said Rockwell Collins had aced its midterm, but the final exam would determine two-thirds of the grade.

"The final exam is still, I think, five to ten years ahead of us. That's when we'll have achieved, at least in my mind's eye, what the full potential of this company is." ▪

Bibliography

Abzug, Barry.
Interview, 2009.

Adelson, James.
Interview, 1983.

Anderson, William G.
Interviews, 1983.

American Aviation,
August 6, 1951.
November 26, 1951.
January 5, 1983.

American Exporter,
February 1947.

Annual reports of the Collins Radio Company, 1945 through 1973.

Annual reports of Rockwell International and Rockwell Collins, 1983 through 2000.

Annual reports of Rockwell Collins, Inc., 2001 through 2008.

"Around the World in 99 Waypoints." *Flying Magazine*, May 1980.

Aviation Week,
December 3, 1951.
August 25, 1952.
September 1, 1952.
November 17, 1952.
March 16, 1953.

Balster, Sherry.
Interview, 2009.

"Battle for Survival." *Forbes*, December 1, 1970.

Bazzell, Cheryl.
Interview, 2009.

Bell, Mike.
Interview, 2009.

Belland, Steve.
Interview, 2009.

Besong, J.P.
Interview, 2009.

Betts, J.A., *High Frequency Communications*. American Elsevier Publishing Company, 1967.

Blocksome, Rod.
Interview, 2009.

Borghese, John.
Interview, 2009.

Bourne, Kenneth M., "HF SSB System Perfected at Collins Liberty Station." *Communications News*, August 1974.

Broadcasting,
April 10, 1950.

Brown, Karen.
Interview, 2009.

Bruce, Jamie.
Interview, 2009.

Bulkeley, William M., "The Fall and Rise: How Rockwell Aided Collins Radio Rebound." *The Wall Street Journal*, March 19, 1973.

"Can It Brew Coffee?" *Rotor & Wing International*, December 1980.

Carlson, Mark.
Interview, 2009.

Cedar Rapids Gazette,
various articles.

Chicago Journal of Commerce,
December 26, 1947.

Chiusano, Bob.
Interview, 2009.

Churchill, Greg.
Interview, 2009.

Churchill, Jim.
Interview, 2009.

Clements, Ralph, *Tales of the Town; Little Known Anecdotes of Life in Cedar Rapids*, Stamats Publishing Company, 1967.

"Climbing Backlog Sparks Gains for Collins Radio." *Barron's*, March 13, 1967.

Collins Column, Collins Radio Company, various articles.

"Collins Radio Has $100 Million Backlog," *Barron's*, February 19, 1951.

Collins Signal, Collins Radio Company, various articles.

"Collins of Collins Radio." *Forbes*, January 15, 1964.

Collins, Arthur A., "Full Details of a Short Wave Transmitter." *Radio Age*, May 1926.

Collins, Arthur A., Speech before the Institute of Radio Engineers, September, 1952.

Colton, Russell.
Interviews, 1983.

Communications,
April 1948. March 1949.

Cosgrove, Jack.
Interview, 2009.

Culver, Senator John, "A Continuing Success Story for Collins Radio." *Congressional Record*, October 6, 1978.

Danek, Ernie, *Tall Corn and High Technology*, Windsor Publications, 1980.

Davenport Democrat,
December 21, 1947.

Des Moines Register,
December 12, 1965.

Dietz, Cindy.
Interview, 2009.

Doelz, M.L., and J.C. Hathaway, "Mechanical Filter."

Dooley, Rod.
Interview, 2009.

Drake, Mike.
Interview, 2009.

Dwhytie, Jeff.
Interview, 2009.

Electronic Industries,
February 1945.

Electronics, July 1940.
October 1951.

Ewoldt, Harold.
Interview, 1983.

FM and Television,
May 1949.

Fischer, Albert, "Flight Cancelled . . ." *Interavia*, Volume IX, No. 4, 1954.

Fishwild, Bruce,
"What Goes On at a Corporation's Meeting?" *The Cedar Rapids Gazette*, November 20, 1963.

Flight, August 1952.
January 1953.

Forbes, January 1968.

Fruehling, Tom, "Rockwell to Add $8.4 Million Building." *Cedar Rapids Gazette*, August 6, 1978.

Gates, Robert S., Speech before the New York Society of Security Analysts, September 15, 1960.

"Genius at Work." *Time*, September 24, 1956.

Gerks, Irvin H.
Interview, 1983.

Girotto, John.
Interview, 2009.

Goetz, John.
Interviews, 1983.

Goodyear, Arlo, *The Collins Story*, October 14, 1954.

Gould, Jack, "New Radio Signal Opens Door to Global Video." *New York Times*, April 30, 1952.

Green, Jim.
Interview, 2009.

"Guglielmo Marconi." *The Wireless Engineer*, September 1937.

Harrison, Gertrude.
Interview, 1983
"He Makes Tinkering Pay." *Business Week*, March 13, 1954.

Hartvigson, Donna.
Interview, 2009.

Henningson, Mary.
Interview, 1983.

Hirvela, Bob.
Interview, 2009.

Hobson, Tom.
Interview, 2009.

Hommer, Don.
Interview, 2009.

Horizons, Rockwell Collins, various articles.

Hoopes, Townsend, "Veteran Collins Pilot Bids Farewell To A Special Lady." *Cedar Rapids Gazette*, May 9, 1983.

Huebsch, Tony.
Interviews, 1983.

Industry Week,
Oct. 1, 1996.

Investor's Reader,
March 4, 1959.

I.R.E. Bulletin, Los Angeles Section, 1953.

Johnson, Dick.
Interview, 2009.

Joliet, Illinois Herald-News,
October 9, 1952.

Jones, Clay.
Interview, 2009.

Kirchenbauer, Ron.
Interview, 2009.

Koomar, John.
Interview, 2009.

Kueter, Dale,
"Anatomy of Economic Turnaround for Collins Radio." *Cedar Rapids, Gazette*, June 17, 1973.

"Collins Name Lives On. " *Cedar Rapids Gazette*, December 4, 1977.

"Collins Wins Major Avionics Contract." *Cedar Rapids Gazette*, July 21, 1979.

Kuster, Joanne,
"Collins Sets Pace with Four Expansions. " *Iowa Development Commission Digest*, December 1977.

Lahr, Millie.
Interview, 1983.

Lander, Clayton.
Interview, 1983.

Larson, Dorothy.
Interview, 1983.

"Latest Search for Northwest Passage: The Plan, the Odds." *U.S. News & World Report*, May 12, 1969.

Latting, Dave.
Interview, 2009.

Lehman, Milton,
This High Man, The Life of Robert H. Goddard. Farrar, Straus and Company, 1963.

Lippisch, George.
Interview, 1983.

Lipses, Karen.
Interview, 2009.

Lueck, Gerrie.
Interview, 2009.

Lund, Merrill.
Interview by Wilma
Shadle, 1980.

Lyon, Evelyn.
Interview, 1983.

Marriner, Ed,
"The Story of Collins Radio."
CQ Magazine, August 1965.

Marvin, Keith. *The Iowa
Engineer*, November 1949.

Mattai, Nan.
Interview, 2009.

McClellan, J. Mac,
"VLF/Omega + Vortac =
Practical Automatic
Navigation." *Business
and Commercial Aviation*,
December 1980.

Metz, Robert, "Collins Versus
the Middle Man." *The New
York Times*, April 24, 1969.

Meyer, Dan, "Advancing
Communications Research
Through an Echo." *Collins
Signal*, No. 4, 1960.

"Microwave Relay Replacing
Cables," *New York Times*.
March 21, 1954.

Millennium, Rockwell
Avionics & Communications,
various articles.

Miller, James.
Interview, 1983.

Myers, Tom, "New Electronics
Firm to Provide Jobs for 300."
Today, June 13, 1976.

Nagle, James J.,
"Personality: An Inventor
Dislikes Publicity." *The New
York Times*, October 1, 1961.

News releases, Collins
Radio Company and Collins
Divisions of Rockwell
International.

Nicholson, Gordon.
Interview, 1983.

Nyquist, John.
Interview, 1983.

O'Connor, Karen.
Interview, 2009.

Opfer, Norman.
Interview, 1983.

Ortberg, Kelly.
Interview, 2009.

Ottumwa Courier,
July 4, 1938.

Ozburn, Jiggs, "The Collins
Radio That I Knew." *World
Radio News*, September 1973.

Pappenfus, E.W., Warren B.
Bruene, and E.O. Schoenike,
*Single Sideband Principles and
Circuits*. McGraw-Hill, 1964.

Pappenfus, E.W.,
"SSB – A Revolution in
Emission. " *Collins Signal*,
Summer 1956.

Pearlstine, Norman,
"Collins Radio Slates
Computer Venture."
The Wall Street Journal,
September 6, 1968.

"Collins Radio Hits Electronic
Data's Take-Over Offer."
The Wall Street Journal,
April 1, 1969.

"Tale of a Tender,"
The Wall Street Journal,
May 8, 1969.

Peters, Leonard.
Interview, 1983.

Pickering, Richard.
Interview, 1983.

Plummer, Rick,
"Comm Central Audio
Visual Presentation."
September 1, 1979. Interview
with Henry Nemec.

Proceedings of the Board
of Directors, Collins Radio
Company, 1933 through 1945.

Prospectus for the Issue
of Common Stock, 1944.

Pulse, Collins Radio Company.

Radio Craft, January 1947.

Radio and Television News,
September 1951.

"Radio." *The New Encyclopedia
Britannica*, 1974.

Reiniger, Scott H.,
"New Picture Aid for ILS –
Omnirange." *Aviation Week*,
July 2, 1951.

Reininga, Herm.
Interviews, 2009.

Rockwell News, Rockwell
International, various articles.

"Rockwell's Surprising
Winner: Collins Radio."
Business Week,
November 15, 1976.

Schleder, Ron.
Interview, 2009.

Schulte, Connie.
Interview, 2009.

Schweighofer, Horst M.,
"Flight Director," *Flug Revue*,
October 1976.

Short Wave Craft,
June-July 1930.

*Signal, Journal of the
Armed Forces Communications
Association*, July-August 1951.